THE MATH TEACHERS KNOW

Education / Mathematics / Concept Study

What sorts of mathematics competencies must teachers have in order to teach the discipline well? This book offers a novel take on the question. Most research is focused on explicit knowledge – that is, on the sorts of insights that might be specified, catalogued, taught, and tested. In contrast, this book focuses on the tacit dimensions of teachers' mathematics knowledge that precede and enable their competencies with formal mathematics. It highlights the complexity of this knowledge and offers strategies to uncover it, analyze it, and re-synthesize it in ways that will make it more available for teaching. Emerging from 10 years of collaborative inquiry with practicing teachers, it is simultaneously informed by the most recent research and anchored to the realities of teachers' lives in classrooms.

Brent Davis is Professor and Distinguished Research Chair in Mathematics Education at the University of Calgary, Canada.

Moshe Renert is a noted mathematics educator in Western Canada with extensive teaching experience at secondary and post-secondary levels.

THE MATH TEACHERS KNOW

PROFOUND UNDERSTANDING OF
EMERGENT MATHEMATICS

by

BRENT DAVIS
MOSHE RENERT

Routledge
Taylor & Francis Group

First published 2014
by Routledge
711 Third Avenue, New York, NY 10017

Simultaneously published in the UK
by Routledge
2 Park Square, Milton Park, Abingdon, Oxon OX14 4RN

Routledge is an imprint of the Taylor & Francis Group, an informa business

Library of Congress Cataloging in Publication Data
    [application submitted]

ISBN: 978-0-415-85843-4 (hbk)
ISBN: 978-0-415-85844-1 (pbk)
ISBN: 978-0-203-79704-4 (ebk)

Typeset in Garamond and Myriad Pro
by Brent Davis

CONTENTS

ACKNOWLEDGMENTS

This book is about understanding the collaborative dimensions of knowing mathematics in a manner that enables and enriches teaching. A central theme is the way in which human knowledge arises in and is distributed across communities, and we thus owe a debt of gratitude to the many, many teachers who have participated in concept studies over the past decade. Particular notes of appreciation are extended to Cynthia Au, Elena Bodnaruk, Elisha Bonnis, Emily Brown, Theo Byleveld, Greg Chan, Pamela Chen, Mike Craig, Lucho Davidov, Rob Deleurme, Karim Dhalla, Brenda Dowle, Tanya Durman, Craig Dwyer, Sara Forsey, Sarah Hamilton, Freddie Irani, Amy Johnston, Chris Krahn, Milly Lin, Debbie Loo, Petra Lungberg, Ed Ma, Cameron MacDonald, Derek Markides, Jennifer Markides, Sandy Miller, Rebecca Mouland, Paula Munoz, Steve Nicholson, Tracey Noble, Shawn Palmer, Kate Pineo, Les Redick, Cheryl Schaub, Sonya Semail, Sheryl Taylor, Amy Tetz, Carla Thomas, Ryan Thompson, Liz Warnick, Brian Wittwer, and Karen Wright.

We would also like to acknowledge and thank Elaine Simmt and Dennis Sumara for their participations in the joint work with teachers that evolved into concept study.

As well, we would like to thank John Mason, Anne Watson, John Mighton, Steven Khan, Lissa D'Amour, Elizabeth Mowat, Katharine Borgen, and Ben Shear for their unique pedagogical contributions to different concept study groups.

TEACHERS' MATHEMATICS

FRAMING THE QUESTION

In brief …

Teachers' disciplinary knowledge of mathematics is one of the most prominent topics of inquiry in the field of mathematics education research at the moment, having first been studied in earnest in the 1970s. We trace the evolutions of this research by tracking the questions (and associated methods, conclusions, and frustrations) that have oriented various studies.

Mathematics-for-Teaching: A Working Definition

[VOICES OF BRENT AND MOSHE]

In the fall of 2009, we visited a local elementary school to discuss student-teacher placements. At the end of that business, the principal invited us into her office, closed the door and, in a whisper, asked if we could offer assistance on another matter.

There were two 1ˢᵗ-grade classrooms in the school, each taught by an experienced educator. Both teachers had strong histories of graduating competent readers who generally enjoyed their schooling experiences. But the situation was different when it came to mathematics.

On paper, these teachers' formal backgrounds in mathematics were almost identical. However, there were marked differences when it came to their teaching. In the one class, students loved the subject; in the other, there was "an epidemic of math phobia" and "scores lower than what we like to see at this school." The principal wondered if we might be able to help.

We agreed, asking that we not be told which teacher was deemed the more effective.

An afternoon "study" ensued. It began with brief, informal staffroom conversations over lunch with the two teachers, during which we were unable to discern significant differences between them. Both educators were passionate about their roles, confident in their mathematics, deeply concerned for their students, committed to their profession, and highly motivated to teach the discipline well. Each could "speak" the language of inquiry learning, manipulative use, knowledge building, meaning making, and multiple representations, with the same relaxed fluency.

To complicate matters, no telltale differences were apparent when we entered the two classrooms. In both settings, the children clearly adored their teachers. Both rooms had the same sorts of mathematical artifacts — number lines, workbooks, counting images. In fact, owing to the school's policy that all classes at the same grade level should stay more-or-less in sync, the lesson topic for both classes was shared: whole number sums less than 20.

But things changed abruptly when the children in the two rooms were told it was time to start math class.

In one room, the announcement invited silence. The lesson was organized around a carefully structured and rehearsed explanation followed by a brief guided discussion. It then moved into focused seatwork on well-sequenced practice exercises in the workbook.

In the other room, the announcement of "It's time for math!" triggered considerable excitement. And while the reasons for this difference were quickly apparent, as we watched we found ourselves struggling to describe what was going on in a manner that would be useful to the principal and to the other teacher. At times, we even wondered whether or not what we were witnessing might appropriately be called a "lesson." There wasn't much of a structured plan to guide the 45-minute time block; there was no frontal instruction, no organized practice, and no pre-stated desirable outcomes. In fact, the only obvious evidence that the teacher had prepared anything was a solitary question that she printed tidily on the whiteboard: "In how many different ways can you add two numbers together to get 15?"

In each of the two cases, then, the students responded in a manner that reflected the lesson's structure. In the first room, they plodded dutifully and predictably through the engineered experience, listening when told to listen, responding when asked to respond, and working when told to work.

Events in the second room weren't quite so orderly or predictable. For instance, students began talking over the teacher before she had finished writing out the question. As the period unfolded, many began to stray off task and worked with sums other than 15. Some used more than two addends to arrive at 15. Some wandered farther afield and included numbers with fractions in their calculations. A lone pair of students in the corner even ventured into signed integers, delighting at the realization that the possibilities were endless when one allowed oneself to use "minus numbers."

It isn't difficult to figure out which teacher's students enjoyed math class more, or why that might have been the case. What may not be so immediately apparent, however, is why that teacher's students outperformed their peers so markedly on achievement tests. The children might have been more engaged and having more fun, but surely the tightly organized and geared-for-the-test experiences of the pupils in the other class should have had some sort of equalizing effect. Why were the test results *so* different?

We maintain that the critical element distinguishing between the two teachers' approaches and their students' performances is to be found not so much in the lessons' structures and delivery (although these were undoubtedly important), but rather in an expertise that precedes structures – namely, the teachers' knowledge of mathematics.

This claim is difficult to defend – not because of the limited data at our disposal, but because the construct of "teachers' disciplinary knowledge of mathematics" is not yet well understood, in spite of decades of intense research. A first stumbling block is illustrated by our opening narrative. The similarity between two teachers' mathematical backgrounds and their formal disciplinary knowledge shows that effective subject matter knowledge of mathematics teachers (*mathematics-for-teaching*, or M_4T, in short) is much more than a readily catalogued or objectively tested set of concepts. M_4T comprises a complex network of understandings, dispositions, and competencies that are not easily named or measured. The embodied complexity of M_4T must be experienced – seen, heard, and felt. The illustrative anecdote is our mechanism for imparting some of that experience within the confines of the written text. We offer many similar narratives throughout this book, in order to illustrate what we mean by the term *mathematics-for-teaching*.

In the events recounted in this opening narrative, for instance, we would contend that entirely different sorts of "mathematics" – that is, two distinct disciplines – were being enacted in the two classrooms. One was about facts and mastery. The other was about engaging with the world through posing questions, identifying patterns, expressing observations, varying the questions, contriving explanations, defending interpretations, and so on.

The differences between the mathematics enacted in the two classes were so vast that we are yet to come to grips with many of them. For example, when we reviewed our notes for the purpose of writing this book, we were surprised to discover that whereas most of the activity in the first classroom was response-oriented, no final answer was given in the second classroom to the original question, *"In how many different ways can you add two numbers together to get 15?"* It is as if the class experienced a collective amnesia about the initial query, allowing it to be forgotten altogether. We believe that's because the engagement for these 6- and 7-year-olds, the activity was not so much about getting the answer. Rather, it seemed to be more about exploring addition and relationships among additive quantities. Through their explorations, the children also engaged in extensive practice with adding and subtracting. According to our estimates, on average, they calculated roughly 5 or 6 times more sums than did their counterparts in the other class.

These important details are strangely easy to overlook. When we ask practicing teachers to comment on this narrative, the topic of teachers' disciplinary knowledge of mathematics rarely comes up. Instead, most often someone begins the discussion with the suggestion that the second teacher was simply using a teaching method that was more suited to this age group. As valid and compelling as this interpretation might be, we regard it as somewhat reductionistic in its separation of *how* one teaches from *what* one teaches. We find it more productive to consider the two classroom events as a contrast in teachers' mathematics knowledge and how this difference is enacted in their teaching styles. The two teachers might have had similar backgrounds in formal mathematics, but to our analysis, their respective understandings of M_4T were worlds apart.

Unfortunately, it is impossible to illustrate this point well without some detailed, moment-to-moment analysis of what was going on in the two classrooms. We did not collect the necessary data in our informal encounter, and have since utterly "contaminated" the research site by involving the two teachers in a larger group of educators who are studying M_4T in their own practices. Their teaching styles have now converged, and they regularly co-plan and combine their classes when math time rolls around. Importantly, this convergence was not prompted by any direct instruction on how to do things differently. Instead, it arose out of a sustained collaborative investigation, which we call "concept study," of what it means to know mathematics. (We describe these sorts of engagements in Chapters 3 to 5.)

To re-emphasize, the sort of mathematics-for-teaching that we are talking about entails much more than "book knowledge." M_4T is, rather, a sort of knowing that is most *real* in action; it inheres in flexible in-the-moment responsiveness. And this is where we offer our working definition of mathematics-for-teaching: *M_4T is a way of being with mathematics knowledge that enables a teacher to structure learning situations, interpret student actions mindfully, and respond flexibly, in ways that enable learners to extend understandings and expand the range of their interpretive possibilities through access to powerful connections and appropriate practice.*

We explain and elaborate this definition as we move through the chapters. We also offer an account of "concept study" as a ready methodology to support the development of M_4T. But before going there, we offer a brief introduction to the broader field of inquiry around the subject of teachers' disciplinary knowledge of mathematics.

What's the Question? Three Framings of Mathematics-for-Teaching

On the surface, the matter of teachers' disciplinary knowledge of mathematics seems fairly straightforward. Deliberate teaching of something – how to sail a boat, how to tie a shoelace, how to factor a quadratic … *anything* – demands a knowledge of whatever is being taught at a level that surpasses the current knowledge of whomever is being taught.

For much of the history of mathematics teaching, this belief was taken as self-evident in schools and universities. Prospective elementary school teachers have typically been required to earn college credits in post-secondary mathematics, and prospective secondary school teachers have often been required to complete university degrees in mathematics prior to enrolling in education programs.

However, the thinking behind these practices is problematic. As Begle (1972, 1979) showed, there is little or no correlation between teachers' college credits in mathematics and their students' performance. Even so, the conviction persists that future mathematics teachers should slog through the same mathematics courses as future physicists, engineers, and computer scientists. Is this a good idea?

Many have argued that it is not.[1] To be honest, we're not so sure as some as our colleagues on the matter. Both of us have studied considerable amounts of mathematics at the university level, and both of us routinely draw on insights gleaned from university courses to inform our teaching at the elementary and secondary school levels. It has been our experience that the time spent in university mathematics courses matters. True, it is not always clear *how* or *why* such study is important or relevant – but that's precisely the point. This is where we align our own thinking with a conclusion drawn by Baumert and colleagues (2010) in a comprehensive review of empirical research on the issue:

> Findings show that [teachers' content knowledge of mathematics] remains inert in the classroom unless accompanied by a rich repertoire of mathematical knowledge and skills relating directly to the curriculum, instruction, and student learning. … In summary, findings suggest that – in mathematics at least – a profound understanding of the subject matter taught is a necessary, but far from sufficient, precondition for providing insightful instruction. (p. 139)

---

1. For more fulsome reviews of these arguments, see Ball & Bass (2003) or Baumert et al. (2010).

One word leapt out at us when we first read these statements – *inert*. The acknowledgement that knowledge of formal mathematics can remain untapped and inactive signals several shifts in thinking around mathematics teachers' disciplinary knowledge since Begle's work. For one, it flags a new focus in the research, whereby the longstanding concern with "what" mathematics teachers should know has been extended to encompass "how" they need to know it. It also acknowledges the many varied facets of the teachers' knowledge.

But we're getting ahead of ourselves.

Our intention in this chapter is not really to answer any of the big questions. Rather, it is to make sense of exactly what questions are being asked. To do this, we offer a chronology of sorts, tracing the evolution of understandings and perspectives among researchers and tracking the questions that have oriented their studies, from the 1970s into the 2010s.

Research into teachers' disciplinary knowledge of mathematics began in earnest in the 1970s, and the already-mentioned work of Begle (1972, 1979) is most often cited as seminal to this branch of inquiry. Begle's studies exemplified a conception of mathematics-for-teaching that dominated research through the 1970s and 1980s, as academics grappled with the question,

What *mathematics* do teachers need to know in order to teach mathematics?	Q1

In addressing this question, the majority of studies of that period sought to identify correlations between two constructs:

- teacher knowledge of mathematics (operationalized in terms of, for example, counts of post-secondary courses in mathematics, grades on those courses, specific course content, and/or performances on standardized tests of formal mathematics);
- student understanding of mathematics (operationalized in terms of, for example, performance on standardized achievement tests, capacities to identify connections among topics, and/or abilities to explain procedures and concepts).

We acknowledge that a great deal of important detail is lost in these brief descriptions. More nuanced accounts of this research are presented elsewhere (see, e.g., Ball, Blömeke, Delaney, & Kaiser, 2012; Ball, Lubienski, & Mewborn, 2001; Baument et al., 2010; Even & Li, 2011; Hill, Rowan, & Ball, 2005; Mewborn, 2001; Roland & Ruthven, 2011; Wilson, Floden, & Ferrini-Mundy, 2002), and we highly recommend those sources to the interested reader. Our purpose here is merely to foreground how the question that prevailed at the time compelled a particular attitude toward inquiry. Specifically, researchers were seeking a relationship between teachers' preparation in advanced mathematics and their students' mathematical learning. Teachers' mathematical knowledge was most often construed in terms of the content of university-level stock mathematics courses. In the decades following Begle's original reports, researchers continued to refine the construct of teachers' mathematical preparation, moving beyond a simple count of post-secondary courses to analysis of the grades the teachers achieved, the instructional approaches taken, and the actual content covered (see, e.g., Monk, 1994).

The persistent lack of significant correlations between teachers' studies in advanced mathematics and their students' performance proved troublesome to researchers, and was a likely contributor to a decline in efforts to investigate the phenomenon in the following two decades. However, there was no apparent decline in the near-universal conviction that knowledge of advanced mathematics was a vital part of teacher preparation. The weak evidence-base in support of this strong conviction was no doubt a principal motivator for a renewed surge of interest in the 1990s.

By then a new way of thinking about teachers' disciplinary knowledge had swept through the field of education: pedagogical content knowledge. Proposed by Shulman (1986), PCK was described as follows:

> ... for the most regularly taught topics in one's subject area, the most useful forms of representation of those ideas, the most powerful analogies, illustrations, examples, explanations, and demonstrations – in a word, the ways of representing and formulating the subject that make it comprehensible to others. (p. 9)

For Shulman, PCK encompassed awareness of both established content and the processes by which content was established:

> It requires understanding the structures of the subject matter. ... The structures of a subject include both the substantive and the syntactic structures. The substantive structures are the variety of ways in which the basic concepts and principles of the discipline are organized to incorporate its facts. The syntactic structure of a discipline is the set of ways in which truth or falsehood, validity or invalidity, are established. (p. 9)

The construct of PCK was provocative and sparked a new interest in mathematics-for-teaching within the English-speaking mathematics education research community. By contrast, the notion was already represented in a significant body of research in several other languages, and was most closely aligned with the European notion of didactiques/didactiks. Worthy of particular mention in this regard in Freudenthal's (1983) *Didactical Phenomenology of Mathematical Structures*, a lengthy and detailed exploration of didactic aspects of mathematical concepts that aligns well with Shulman's more general formulation of PCK.

During the 1990s, an increasing number of studies picked up on PCK as a means to reframe the mathematics-for-teaching question. Prominent among them was Ma's (1999) examination of contrasts between the content knowledge of Chinese and American teachers. She provided anecdotal evidence of highly specialized knowledge of elementary mathematics that manifested prominently in the practice of Chinese teachers, but not in that of American teachers. Ma used the term *profound understanding of fundamental mathematics* (PUFM) to refer to this knowledge. She explained:

> A teacher with PUFM is aware of the "simple but powerful" basic ideas of mathematics and tends to re-visit and reinforce them. He or she has a fundamental understanding of the whole elementary mathematics curriculum, thus is ready to exploit an opportunity to review concepts that students have previously studied or to lay the groundwork for a concept to be studied later. (p. 124)

According to Ma, PUFM comprises four key components – connectedness, multiple perspectives, basic ideas, and longitudinal coherence among the concepts that comprise grade-school mathematics curricula.

Ma's study helped to trigger a rethinking of the orienting question of M_4T among researchers, and efforts to identify the specialized types of teachers' disciplinary knowledge dominated M_4T research in the first decade of the 21st century. In that time, the orienting question for researchers shifted from, "What *mathematics* do teachers need to know in order to teach mathematics?" toward,

What *specialized mathematics* (i.e., PCK) do teachers need to know in order to teach mathematics?[2]	Q2

It is important to emphasize that the earlier question was not lost in this revised formulation. The shifting focus was never intended to suggest that courses in formal mathematics do not matter, but rather to signal other categories of know-how that up to that point had not been given the airtime they merited. As Monk (1994) phrased it, "a good grasp of one's subject areas is a necessary but not a sufficient condition for effective teaching" (p. 142).

---

2. Ball and colleagues (e.g., Ball, Thames, & Phelps, 2008) identify several subcomponents to teachers' disciplinary knowledge of mathematics, distinguishing among for example "specialized content knowledge" and "pedagogical content knowledge" (see Chapter 5). We use the term "specialized mathematics" to encompass all components of disciplinary knowledge pertaining to teaching, acknowledging that much of the subtlety of their typography is omitted here. Our point is simply that a critical evolution in the field was the recognition that teaching entails a specialized disciplinary knowledge that is distinct from what might be gleaned from more advanced study.

There are at least two ways of interpreting Monk's statement – either as a call for *additional knowledge,* or as a call for *knowing differently.* The field's first impulse was to move with the former, a shift in sensibility that might be represented in terms of nested phenomena:

Consonant with the suggestion that additional knowledge is entailed in the work of teaching mathematics, others have pointed to further dimensions of content knowledge. Kilpatrick, Swafford, and Findell (2001), for example, identified "five strands" of teachers' content knowledge: conceptual under-standing, procedural fluency, strategic competence, adaptive reasoning, and productive disposition. Moving in a different direction, Fauvel and van Maanen (2000) argued that awareness of the history and epistemology of mathematics is a vital component of teachers' content knowledge. The list goes on.

The addition of PCK to the M_4T mix highlighted an important distinction between *mathematical knowledge that is structured to be used* (by mathematicians, physicists, engineers, etc., and ultimately by students) and *mathematical knowledge that is structured to be taught* (i.e., used by teachers in the didactic process). This important recognition highlights a unique aspect of teachers' disciplinary work: they must translate formal and utilitarian knowledge into learnable knowledge (or, in terms proposed by Brousseau, 1997, engage in "didactical transformation").[3]

In spite of the new focus, useful answers remained elusive. In particular, it was not clear what set of specialized concepts and results could constitute a body of mathematics useful for teaching. Oriented by this issue, Ball, Thames, & Phelps (2005) suggested a new vantage point, by pointing out that the mathematical knowledge of teachers is not static and should be thought of as knowledge-in-action. In their view, mathematics teaching is a form of mathematical practice that includes, among other elements:

---

3. It's also important not to overstate the distinction between these categories of mathematics knowledge. Brunner et al. (2006) have shown a high correlation between them.

- interpretation and evaluation of students' work;
- correlation of students' mathematical results with the processes of their production;
- construction of meaningful explanations, and assessment of curriculum materials.

Ball and colleagues called for a practice-based theory of teachers' mathematics knowledge that focuses on the specific knowledge that teachers use in their daily work. The framing question for the study of M_4T evolved more toward,

> What mathematical knowledge *is entailed by the work* of teaching mathematics?"
>
> Q3

Once again, this formulation involved an elaboration of earlier foci and not a rejection. The new formulation might be characterized as another layer in an expanding sphere of knowledge:

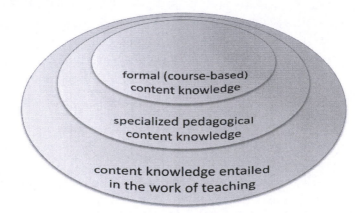

formal (course-based) content knowledge

specialized pedagogical content knowledge

content knowledge entailed in the work of teaching

In contrast to the second iteration of the M_4T question, which turned attentions toward *knowing more*, this third iteration also gestured toward *knowing differently*. The shift is clearly illustrated by Ball and Bass's (2000, 2003) research, which focused attention on a key process of teachers' mathematical practice that they, following Ma (1999) called *unpacking*. Unpacking is the prying apart and explicating of mathematical ideas to make sense of their constituent images, analogies, and metaphors. Whereas mathematicians often convert their ideas into highly condensed representations to facilitate mathematical manipulation, teachers employ the reverse process of unpacking ideas to reveal and explain the meanings of mathematical constructs. Adler and Davis (2006) have also studied unpacking, along with other aspects of teachers' mathematical work. Their somewhat worrisome findings indicated that mathematical ideas addressed in teacher education courses in South Africa are predominantly compressed, not unpacked.

Returning to our assertion that much of the current research into M_4T in the English-speaking world was anticipated by the European didactiques/didactiks literature, we note that Freudenthal's didactical phenomenology extended beyond a study of mathematical structures, and examined mathematical "objects" as teachable and learnable forms. This didactic emphasis on mathematical knowledge entailed:

> … knowledge of mathematics, its applications, and its history. … how mathematical ideas have come or could have come into being … how didacticians judge that they can support the development of such ideas in the minds of learners … understanding a bit about the actual processes of the constitution of mathematical structures and the attainment of mathematical concepts. (p. 29)

We can readily see that as one delves more deeply into the M_4T question, one finds ever more nuances, layers, and enactments of mathematics.

To recap, as this brief overview of M_4T shows, three different research questions have oriented the field since the 1970s:

Q1) What *mathematics* do teachers need to know in order to teach mathematics?
Q2) What *specialized mathematics* (i.e., PCK) do teachers need to know in order to teach mathematics? and
Q3) What mathematical knowledge *is entailed by the work* of teaching mathematics?

Correspondingly, three key answers have been offered to the question of what constitutes mathematical knowledge for teaching. They are:

A1) teachers need to know more advanced mathematics than the mathematics they are teaching;
A2) teachers need to know specialized mathematics (i.e., PCK); and
A3) teachers' mathematical knowledge is enacted in their daily work and involves unpacking/decompressing of content.

Again, it is important to recognize that successive questions do not obviate earlier concerns. Rather, the second iteration of research encompassed and elaborated the first iteration, just as the third iteration included and transcended the previous two. All of them matter.

We strive for this same attitude as we move toward our own formulation of the question, which we see as another layer on the nested M_4T image. To that end, we repeat our "answer":

A4) *M_4T is a way of being with mathematics knowledge that enables a teacher to* *structure learning situations, interpret student actions mindfully, and respond flexibly, in ways that enable learners to extend understandings and expand the range of their interpretive possibilities through access to powerful connections and appropriate practice.*

Of course, we haven't yet posed our question. To do that, we need to situate our work in current discussions of knowing and learning, which is where we go in Chapter 2 .

Toward that end, it is also important to flag the issue of how where and how a knowledge of mathematics-for-teaching is learned. For Begle, the assumption was that it was developed in mathematics courses, particularly those taken in post-secondary settings. As is clear from the way Q3, above, is posed, more recent research leans toward the assumption that teachers' disciplinary knowledge is highly situated. Hence it is largely developed and learned in practice. In this writing, we join with a few others (e.g., Rowland & Ruthven, 2011; Zazkis, 2011) in embracing both insights. However, we also argue for the critical importance of a third "site" – namely, the community of teachers working together to understand their mathematics for the purposes of teaching it.

The Bigger Picture: Making Sense of M_4T by Grappling with Why Math Matters

The balance of this book is about how our definition of M_4T embraces, critiques, and elaborates other perspectives. As a starting place, in the chapters that follow we offer a conception of M_4T that adopts a more critical stance to several issues that we feel are un- or under-addressed by the three questions and responses discussed above. In particular, we believe that the perspectives represented through Q1, Q2, and Q3, share common assumptions and a tacit attitude toward these four elements:

- context – all are constrained in scope to the immediate worlds of teachers and their students;
- content – none incorporates a critical attitude toward school mathematics curricula as a central aspect;
- control – each is readily aligned with entrenched conceptions of the teacher as a person who manages the relationship between a more-or-less fixed body or knowledge and an unfinished knower; and
- connectedness – in every case, it appears that the unit of analysis – that is, the "learner" – is an individual person (whether teacher or student) rather than, for example, the collective knower that arises in the connected minds of individuals or the body of M_4T knowledge.

It's going to take us the entire book to even begin to deconstruct these (and other) deeply entrenched assumptions about *context, content, control,* and *connectedness* and, in the process, to elaborate our definition of mathematics-for-teaching and attempt to capture the subtle complexity of the phenomenon. Nevertheless, we end this first chapter by laying out some of the explicit beliefs that we bring to these matters.

On the matter of **context**, a worrisome criticism of modern schooling is that the institution is increasingly out of step with our rapidly evolving society. Critique after critique reminds us that the modern school maintains structures and emphases that were developed to meet the needs of the industrial era. It aims to equip learners with skills that are of decreasing value in a knowledge-based culture in which existence is becoming subtly but pervasively digitized (see, e.g., Davidson, 2011). In the new economy, mathematics has emerged as *the* gateway discipline – that is, the *numerati* have supplanted the *literati* in access and influence (Baker, 2008). Yet the sorts of mathematical competence that are of value to current and future society bear ever-diminishing resemblance to the emphases seen in contemporary classrooms. Schooling can no longer be construed as direct and relevant preparation for the labor market, as many – perhaps most – of the careers that will eventually be taken on by today's children simply have not been invented yet.

Mathematics educators are well aware of the situation, but in spite of large-scale initiatives to rethink the **content** of school mathematics (e.g., NCTM, 1989, 2000), so far, few educational institutions have demonstrated a capacity to match the pace of cultural change. It is true that new technologies (e.g., graphing calculators, computer algebra systems, dynamic interfaces) are replacing old ones (e.g., pencil-and-paper arithmetic and algebra), enabling students to solve more sophisticated and realistic problems than had traditionally been the case. However, the fundamental content has changed very little from previous decades and even centuries. Many reasons can be (and have been) mentioned: social inertia, limited resourcing of schools, fatigue with ever-swinging reform pendula, systemic resistance to reform, and so on. We believe that a critical (if not the main) issue here is a lack of capacity in teachers to deal with new mathematical content, and even with current content, in compelling ways. In other words, the stagnation has much to do with teachers' disciplinary knowledge of mathematics.

In advocating greater teacher flexibility with mathematics, we do not mean to suggest or even imply that every teacher should possess a broad and nuanced knowledge of everything taught. We are actually going to argue in the pages that follow that this sort of expert knowledge is nearly impossible. What is achievable, we believe, is *a shift in how teachers deal with what they do not know yet.* This shift entails a very different way of thinking about who is in **control** of what in the classroom. In a nutshell, the role of mathematics teachers must change. They can no longer be positioned as trustees, managers and arbiters of sanctioned knowledge. Rather, teaching has to be seen as a co-participation in the production of knowledge, among teachers, and with their students. Importantly, this is more about a shift in perception than a shift in formal knowledge.

As we attempt to demonstrate, teachers and their students have always been participants in the production of mathematical knowledge, but not necessarily in the explicit ways that are often associated with formal mathematical knowledge. Rather, their influence manifests in a far more subtle and pervasive way than is commonly noted. We can better understand it by learning to appreciate the **connectedness** that pervades mathematics and its pedagogy – connectedness of ideas within a mathematical domain and among domains, connectedness of knowledge and knowers, connectedness of knowers to one another, and connectedness of humans to the more-than-human world. When the language of Western schooling is analyzed, more often than not it is rife with notions of breaks, discontinuities, dichotomies, and dyads. Teacher is cast against student, self against other, individual against collective, subjective against objective, nature against nurture, humanity against the rest of the world, and the list goes on. Yet something very powerful happens when we shift attention to how things are linked rather than how they are distinct. Specifically, with regard to the major issues that are confronting society and humanity today, we join with many others in the hope that current interest in connectedness holds a key to finding and inventing effective and appropriate responses. We see connectedness as a means to ongoing viability. Even something as mundane as teachers' disciplinary knowledge of mathematics can make an important contribution, as it is necessarily connected to many facets of human experience.

Let us return to our opening narrative to highlight some the key points of this section. The members of the second, livelier class were a **connected** learning community. They were connected both to one another and to the emergent ideas that arose in their intersecting inquiries. These connections arose because the teacher permitted so much of what went on in the room to be out of her **control**. The result was a mathematical **content** that was so rich and varied that it easily surpassed anything specified in curriculum guides, teachers' manuals, and student textbooks. It was content that embodied an adaptable flexibility, mirroring the grander social, cultural, and global **contexts** of the learning.

And it was all about the teacher's disciplinary knowledge of mathematics, that is, the teacher's M_4T.

Where Are We Going with This?

A few paragraphs ago we listed some of the dyads that have been prominent in discussions of schooling. We will be deconstructing all of them in one way or another in later chapters, as we regard such critical examination a precondition for opening new possibilities for making sense of M_4T.

By way of illustration, two of the dyads examined in Chapter 2 are explicit *vs.* tacit knowledge (Polanyi, 1966) and reflective *vs.* automatic knowing systems (Kahneman, 2011). These pairings are particularly useful in making sense of the differences between the two enacted versions of school mathematics introduced at the start of this chapter. Our preliminary interviews showed that the two teachers were on very similar footings when it came to explicit knowledge of mathematics and reflective thinking on matters of teaching. But life in the classroom revealed something very different. These teachers diverged on matters of tacit knowledge and automatic thought, two vital elements of M_4T.

Let us extend the discussion by moving to a different anecdote in a different classroom – another window through which we can catch sight of a more complete M_4T.

KNOWING AND LEARNING (MATHEMATICS)

SOME GAME-CHANGING INSIGHTS

In brief ...

This chapter situates our version of the M_4T question. We focus on a handful of game-changing developments in the cognitive and complexity sciences. The chapter comprises a series of brief accounts of relevant recent constructs that inform current discussions of knowing and learning, generally, and the question of teachers' disciplinary knowledge of mathematics, specifically.

"Thou Shalt Not" – A Teacher Dares to Divide by Zero

[SPEAKER: MIDDLE-SCHOOL TEACHER, TASHA]

In 2007, the mathematics curriculum in British Columbia was changed. While reviewing the new document, I noticed an addition to the standard 7th-grade learning outcome that covers the divisibility rules for 2, 3, 4, 5, 6, 8, 9, and 10: explain "… why a number cannot be divided by zero."

My first response was, "That's new." My second was, "Wait, I don't have any idea why you can't divide by zero! How can I teach that?"

In my first attempt teaching the concept I relied on grade-level resources, which recommended an "exploratory" lesson. I did this after a warm-up intended to highlight the relationship between division and the other operations (particularly subtraction and multiplication). Reading student work afterwards, I realized that the lesson had fallen short in a fundamental way: students didn't differentiate between the expressions $0 \div 8$ and $8 \div 0$. Even though they gave an answer of "zero" for the first and an answer of "no answer," or "undefined," for the second, students concluded that both of these responses meant "nothing," and therefore the two questions were the same. This led to re-teaching with explicit instruction: $0 \div 8 = 0$, which has an answer; $8 \div 0$ is undefined, which means it has no *answer, and is* not *possible. It is* not *the same question as $0 \div 8$.*

The following year, for my second attempt, I went back to the teachers' guide and found more explicit background information using models of repeated subtraction and inverse multiplication. After the failure of the exploratory lesson, I tried a more structured approach, introducing the models to the students first with possible divisions, then with division by zero. This time students were able to distinguish between the different questions, but their explanations tended to ignore the models provided (on repeated subtraction and inverse of multiplication) for their own responses based on grouping. These explanations were fuzzy, vague, and often incorrect; students were able to state that a number could not be divided by zero, but the class could not explain why.

I sought out several grade-level colleagues for advice on how they taught the lesson. Their responses? They didn't – they skipped the concept entirely. I asked some high school teachers how they respond to student questions on division by zero. Their responses? "Don't. Just don't."

My third attempt at teaching the concept began with the question, "What is division?" Students worked in groups to brainstorm models of division, which were brought together on chart paper as a class. Collectively, students developed models based on grouping, repeated subtraction, number-line hopping (including the goal being zero), fractions and the inverse of multiplication. They then used the models to specifically model the equation $6 \div 2 = 3$. Next, the students were asked to use the models to represent $6 \div 1 = 6$, and $0 \div 6 = 0$. I then changed the question to $6 \div 0 = ?$, and asked students to work with the models. Students began to respond with phrases such as, "It doesn't make sense" and "It's all wonky" – which led to introduction of the term "undefined." In follow-up work, the success was clear: students could answer the questions about division by zero correctly, and also clearly explain why, using the meaning of division. Student work showed several different models, explained the difference between dividing by 0 and 1, and some even grappled with the ideas that it "suggests infinity" while being not able to provide a numerical answer, and was thus undefined.

Tasha is a middle school teacher who took part in a two-year master's program focused on teachers' disciplinary knowledge of mathematics. That program, offered through the University of British Columbia, began in the summer of 2009 – just before Tasha's third attempt at teaching what it means to divide by zero. In the final paragraph of the above narrative, reference is made to some teaching strategies that Tasha adapted from her M_4T master's work. In Chapters 3 and 4, we delve into these strategies and their origins in more explicit detail as we introduce "concept study."

Like most other examples that we present in this book, Tasha's narrative is offered as research data in support of the assertion that a way of knowing mathematics *exists* that can have powerful influences on teaching. The nature of such knowing, however, isn't just a matter of mastering formal mathematical concepts. After all, teachers' primary concern is not with *mathematics*, but with *learning mathematics*. Some of the considerations entailed in this distinction are the foci of this chapter.

Such distinctions in education aren't easy to broach. Many theories of knowing and learning have risen to prominence in the field of mathematics education. More often than not, they're discussed in terms of their superficial differences rather than their deep compatibilities. In wishing to avoid getting caught in the mire of current debates, and attempting to sidestep the need to align the work with any single theory of learning, we take a different tack. By focusing on common ground of varied theories and by avoiding self-classifying ourselves, we aim to steer clear of questions of "Why is this one right and those are wrong?" and to take on the deeper question of "In what ways is each correct?"

Our reason for beginning a discussion of teaching by explicating these principles of knowing and learning is straightforward. There are hundreds of synonyms of and metaphors for *teaching* in English (Davis, 2004), and the list includes some patent opposites, such as *educating* (literally: drawing out) and *inducting* (literally, drawing in), and *facilitating* (i.e., making easy) and *challenging* (i.e, making difficult). Indeed, across the many, many interpretations of teaching that one encounters, there seems to be only one point of agreement: teaching has to do with prompting some sort of learning.

What, then, is *learning?* Just as with conceptions of teaching, as one delves into the everyday language and the formal literature, one finds a stunning diversity of perspectives on learning. Once again, they seem to cluster around a single point of agreement: learning has to do with affecting knowing.

And that is why, in this discussion of teaching, we begin by being explicit, to the extent we are able, on some of the principles of knowing and learning that frame our discussion. To do this, we call on recent key findings in the cognitive sciences, ones that are informed by prior developments in complexity science, which is the topic addressed in the next section,

In terms of necessary background knowledge for the balance of the book, we see the next section (on complexity) as crucial. Subsequent sections are highly relevant, but skimmable for the time being. We actually toyed with the thought of pushing them aside into an appendix, but ultimately we agreed that it is important to be clear on the point that our version of the M_4T question doesn't come out of nowhere. The details provided in this chapter offer a glimpse into why we ended Chapter 1 by highlighting our concerns with the manners in which matters of context, content, control, and connectedness are addressed (or, more often, not addressed) in the bulk of current research into teachers' disciplinary knowledge of mathematics.

Complexity Science

Over the past several decades, the word *complexity* has taken on a very specific meaning in most academic domains. Unfortunately, the shift in meaning has yet to register in the field of education in the way it has in economics, business, physics ... and most other research areas. For that reason, we proceed with a few framing remarks on our understanding of the notion, endeavoring to pry its meaning away from the commonsense interpretation of "really complicated." Complexity is something entirely different.

The shift in meaning is rooted in complexity science, which is an emergent academic movement that has only cohered over the past half century. It originated in the physical and information sciences, but its interpretations and insights were quickly brought to bear in an increasing range of social areas in recent decades. To a much lesser (but noticeably accelerating) extent, complexity has been embraced by educationists whose interests extend across such levels of phenomena as genomics, neurological process, subjective understanding, interpersonal dynamics, cultural evolution, and global ecology (Davis & Sumara, 2006).

Complexity science is itself an example of what it studies: an emergent phenomenon in which similar but nonetheless diverse elements coalesce into a coherent, discernible unity that cannot be reduced to the sum of its constituents. A sense of its internal diversity might be gleaned from some of the varied terms for complex systems that arose in different domains, including "complex adaptive systems" (physics), "nonlinear dynamical systems" (mathematics), "dissipative structures" (chemistry), "autopoietic structures' (biology), healthy organisms (health care), "organized complexities" (information science), and simply "systems" (cybernetics) (Mitchell, 2009). On a less academic plane, the following are among the most frequently noted exemplars of complex systems/phenomena: ant hills, brains, cities, economies, cultures, and ecosystems.

These ranges of titles, interests, and entities are both boon and bane. On the positive side, they offer a sense of the breadth of phenomena and diversity of interests that are addressed within discussions of complexity. On the negative, they make it impossible to offer a readily comprehensible, one-size-fits-all description of complexity. To this end, prominent among efforts at a coherent, unified definition are such terms as *emergent, self-organizing, context-sensitive*, and *adaptive* (Johnson, 2001). As educators and educational researchers, we find a particular resonance with the notion that a complex system is a system that *knows* (i.e, perceives, acts, engages, interprets, etc.) and *learns* (adapts, evolves, maintains self-coherence, etc.). This "definition" figures prominently in this writing. (See Davis, Sumara, & Luce-Kapler, 2008, for a more nuanced discussion.)

One visual metaphor that we find particularly useful in making sense of the scope of foci of complexity research is an image of nested ecosystems (see Figure 2.1). While far from comprehensive, this image highlights some of the nested ecosystems that are of prominent interest to educators and educational researchers – from the subpersonal (the endo-ecosystem, comprising the brain and other bodily subsystems), the personal (ego-), social and institutional (edu-), cultural (ethno-), and ecological (enviro-). These inextricable ecosystems unfold from and are enfolded in one another, each simultaneously maintaining its own proper coherence while playing an integral role in a larger ecosystem. In a very real sense, it is impossible to study one without studying all of the others. At the same time, with each new level of emergence, new laws arise that cannot be anticipated or explained strictly by reference to prior systems. That is, each is incompressible. To understand any level of a complex ecosystem, it must be studied in its own right.

This particular detail is highlighted in the "Some Studies of Knowing and Learning" column in Figure 2.1. Viewed through the lens of complexity, disciplines as diverse as neurology and anthropology are understood as methodologically robust ways to study dynamic, complex ecosystems – learners. In no way can these domains (or the associated learning systems)

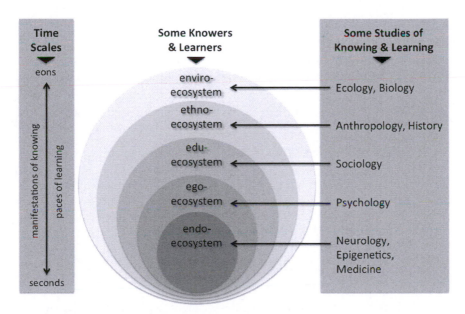

Figure 2.1. Some of the complex nested systems of interest to educators.

be reduced to or construed in terms of one another. At the same time, in no way can they be treated as independent and self-contained. Complexity thus invites a different attitude toward the discursive diversity one finds among domains of study. As we endeavor to demonstrate with the example of M_4T, more might be gained by bringing varied discourses into conversation than positioning them as distinct (or, more troubling, as being in tension).

Phrased differently, we believe that mathematics teachers' disciplinary knowledge is most productively construed as a complex phenomenon – an assertion that aligns with Foote's (2007) observation that mathematics is itself a complex system that is approaching the limits of human verifiability:

> the human element is an essential agent in the evolution of mathematics as a complex system, the the "layers" of complexity mirror the "knowledge states" in this adaptive process. (p. 412)

Like mathematics, then, M_4T comprises multiple co-implicated, emergent, and evolving ecosystems. Numerous systems of knowing and learning – spanning multiple levels of organization – are all at play when we consider what constitutes M_4T. By acknowledging the complex nature of the phenomenon, we can sidestep reductionistic attempts at description, and instead focus on fruitful ways to condition evolution of teachers' knowledge.

With this complexivist frame in mind, we move now to our discussions of knowing and learning. Our strategy for examining these constructs is to look at some of the dyads – that is, distinctions and discernments – that are embraced across a diversity of educational theories. Continuing with the theme of "transcend and include" introduced in Chapter 1, we offer that it is more generative to view these pairings as complementarities, and not as dichotomies. That is, they are co-emergent and productively interdependent. More colloquially, rather than dwell on either/or tensions, we are aiming for both/and syntheses, and we borrow the logical operator "v" (i.e., the inclusive "or") from mathematics to signal this attitude.

We thus move through the following pairings:

- knowing dyads: expert **v** novice
  fast **v** slow
  explict **v** tacit

- learning dyads: logical **v** analogical
  surface **v** deep
  individual **v** collective

For us, the operator "v" actually does more than name the operation of uniting. With its visual similarity to the Roman "v," it also reminds us of how easy it is to frame the elements in these sorts of pairings in terms of one *versus* the other.

This is not a trivial point. The more we grapple with the question of M_4T, the more we confront the necessary simultaneity of teachers seeing themselves as, for example, experts *and* novices. It's not a matter of one or the other, or even of relative weighting. It is about the complex harmonies that emerge in the interplay of these sorts of pairings.

Knowing

In discussions of education over the past 30 years, the word *knowing* has slowly but steadily been taken up in place of *knowledge*. There are many reasons for this shift in preference. For example, *knowing* signals an inseparability of knowledge and knower. That is, the shift in vocabulary flags awareness that it makes no sense to talk about something known without also talking about who or what knows it. (Some philosophers of knowledge might debate this point, but it is difficult to dispute it in an educational frame.)

Further, as a gerund, *knowing* reminds us of the dynamic characters of both knowers and knowledge. These are ever-changing, evolving forms. The word is also intended as a reminder that knowing is contextual and embedded, and the contexts of knowing are similarly dynamic.

To get to the punch line, we regard M_4T as a knowing system, and this rhetorical move compels us to be clear about the interactive, co-implicated natures of the knowers, knowledge, and their contexts that are entailed when "knowing M_4T" is invoked. That is a major part of the project of the remainder of the book. For now, we review some recent thinking around three dyads that others have researched to inform understandings of knowing.

Knowing: expert v novice

As recounted in the opening narrative, Tasha's third attempt at the concept of division by zero began with the query, "What is division?"

As innocent and straightforward as this question sounds, it can be a difficult one for the expert knower to answer in explicit terms. That's because there are many possible responses – division can be defined in many ways – and so, while experts usually don't need to think too deeply to recognize when something is about division, they may not have conscious access to the reasons their ready interpretation.

In our own work with pre-service and practicing teachers, we've attempted to illustrate this point by offering lists of different situations, such as the following:

- dealing out a group of objects into equal-sized sets;
- making groups of a preset size from a group of objects;
- making hops of pre-determined length along a number line; and

- splitting a span on a number line into a predetermined number of segments.

Almost without fail, audience members recognize all four of these operations as instances of division, even though the underlying physical actions (i.e., distributing, clustering, hopping, and folding – respectively) are very, very different. Experts move fluidly among these (and other) instantiations of division, not having to consider their differences or to contemplate which underlying action (if any) fits with whatever situation is encountered. That is because, for experts on division, conceptions are complex blends of the sorts of instantiations listed above – a point that is demonstrated a frequent inability to perceive significant differences between such instantiations. Indeed, more than one of our audience members has fired back some variation of, "They're the same thing!" or "There's no difference between those examples!"

That manner of response is really a variation of the oft-noted observation that, when confronting a new problem or a novel situation, the novice sees many possibilities and the expert sees only a few. Having had opportunity to meld diversities of experience into coherent interpretations, experts are experts precisely because they don't have to consciously sort through how things might be similar and how they might be different. For an expert on division, an instance of division is recognized as an instance of division. Full stop.

Unless, of course, that expert is a teacher. As noted in Chapter 1, a major dimension of the teacher's work is to translate utilitarian knowledge into learnable knowledge. In other words, the teacher must be consciously aware of the many interpretations of a concept that might be circulating in a classroom. This insight has been well represented in the school mathematics literature for several decades now. Manuals for pre-service teachers and textbooks for classrooms are rife with multiple interpretations of basic operations and other concepts. Such explicit representations aren't enough, however, as Tasha's second attempt at division by zero illustrates. Teachers need to be able to do more than to unpack a concept or to pry apart various instantiations. They also have to be able to provide learners with the expert assistance needed to blend instantiations into a coherent concept. In other words, the expertise associated with *using* mathematics is quite different from the expertise associated with *teaching* mathematics. In the vocabulary of expert v novice, *teachers are experts who are able to think like novices.*

There is much more to be said on the differences between experts and novices, and we'll be revisiting the topic in subsequent chapters. For now, we'll toss in that studies of experts and novices have a long history in the mathematics education research literature (see, e.g., Leinhardt, 1989;

Schoenfeld & Herrmann, 1982) and a very deep history in psychology and related domains (e.g., Chase & Simon, 1973; Ericsson et al., 2006). Among the important findings are that, within their domains of strength, experts are able to remember more, remember quicker, and deal more effectively with unfamiliar material as they engage in more effective and efficient ("best") strategies for sorting through, decomposing, and formatting information.

Knowing: fast v slow

Where does the blazing speed of the expert come from?

One explanation that has gained considerable traction over recent decades is based on a "dual-process model" of the brain. As Kahneman (2011) explains, humans apprehend the world in two radically different but simultaneous ways. That is, humans use two fundamentally different modes of thought, which Kahneman labels "System 1" and "System 2."

System 1 is fast – mostly because it is memory-based and seemingly automatic. When called to act, it draws on practiced skills, rehearsed lines, and experienced situations. System 1 is analogical – associative, metaphorical, impressionistic. In operation it can feel like intuition, given its capacities to take on great amounts of information and present us with immediate interpretations that are relevant and accurate. And we don't have any intentional control over it.

In contrast, System 2 is slow – mostly because it is thinking based. It is what we rely on when confronting a situation that is overly unfamiliar or confusing – that is, when we have no practiced procedures or pre-rehearsed interpretations to call on. System 2 is deliberate and requires effort, and is therefore slow and easily tired. And assuming that there are no major problems with the brain's executive functions, System 2 can more-or-less take responsibility for initiating and persisting with the task at hand.

The level of control that we have over our System 2 seems to support a common and persistent delusion. As Kahneman describes it, System 2 likes to think of itself as the lead actor but is really a supporting character. In fact, most often, System 2 is happy to hand over control to System 1 – which usually turns out to be an appropriate thing to do. System 1 is highly attuned to subtle environmental cues, usually affords suitable responses, and frees up System 2 to kick into gear when it must.

In other words, the expert has educated her or his System 1, which is what enables her or his System 2. Importantly, this is not an either/or situation. The two systems can and do amplify one another's capacities. With regard to mathematical understanding, most adults who have attempted to help young children make sense of basic operations will have experienced the effects of this co-amplification as they wondered why they ever had difficulty with, for example, lining up decimals, rounding numbers, or adding simple fractions. With an adequate depth of experience and the time to weave it all together, the facility that System 1 can achieve has the potential of obscuring the actual complexity of coming to understand.

As far as contempory research of M_4T goes, most studies to date have focused on System 2: conscious, thinking-based, slow knowing. The balance of this book might be appropriately decribed as an argument for a more harmonious attention to the automatized knowing that one might not "know" one knows: System 1.

Knowing: explicit v tacit

We would be remiss if we failed to mention the pioneering work of Michael Polanyi (1966) as it relates to the topics just discussed. In many ways, he anticipated the above dyads in his examinations of the explicit and tacit dimensions of personal understanding. For him, explicit knowledge is that subcategory of human awareness that has been codified – that is that can be expressed symbolically (e.g., in words or numbers) and readily shared (verbally, in writing, through libraries, online, etc.). Such knowledge has been the major focus of the modern school, in large part because it seems objective and is easily tested. One either is explicitly aware of something or is not.

Tacit knowledge, in contrast, is highly personal and can be hard to symbolize. That makes it difficult to share with others. More fundamentally, however, it also makes it more difficult to share it with oneself – that is, to bring it to explicit awareness. Tacit knowledge is simply knitted into one's being – enacted, embodied, performed, taken-for-granted. This category thus encompasses personal insights, intuitions, hunches, convictions, values, and morals.

Just as with the other dyads discussed in this chapter, the pairing of explicit and tacit knowledge is not intended as a dichotomy. The categories are mutually dependent. Our interest in the dyad here is with regard to the manner in which formal education has overwhelmingly focused on explicit know-how, just as its implicit emphases have been on logical knowledge, expert knowers, and slow (System 2) knowing. Certainly such foci prevail in current research into teachers' disciplinary knowledge of mathematics. Our argument is that such emphases are important, but patently inadequate. Yes, mathematics teachers must have logical knowledge, be expert knowers, be capable at slow knowing, and have explicit know-how. However, all of these emergent capacities are enabled by categories of knowing that are much harder to notice and, consequently, much harder to teach.

We'd like to make the claim that our own efforts to understand M_4T have not fallen into the trap of over-emphasizing explicit knowledge and under-addressing tacit knowledge. Such a claim, however, would be somewhat inaccurate. Two earlier, and very unproductive, attempts to investigate teachers' disciplinary knowledge made precisely this mistake. The first was an effort to develop a formal interview protocol that might be used to assess the range of instantiations of core concepts available to a teacher – by which we did little more than demonstrate that teachers' capacities to explicitly identify such instantiations paled against their capacities to tacitly invoke them in situations requiring application or interpretation. Spurred by that disappointing effort, but encouraged by the realization that teachers did in fact know more than was available to conscious awareness, we attempted to develop protocols for classroom observations. The results were just as disappointing, but in a different way. Teachers did indeed demonstrate a broader knowledge of instantiations as they worked with students to make sense of various concepts. For instance, in a fourth grade lesson on division, it wasn't unusual to observe reference to all four instantiations for division mentioned above (distributing, clustering, hopping, and folding). However, teachers' uses of these instantiations were rarely conscious and systematic. Rather, diverse interpretations of core concepts were typically invoked opportunistically. They were rarely considered in conjunction with other instantiations, even more rarely explored in any depth, and never with any apparent awareness of the fact that novices might not be able to reconcile instantiations that the expert can no longer see as different (Sfard, 2008).

Of course, all of this just begs the question: How do we approach the teaching of M_4T if its vital aspects aren't so readily available to consciousness? We don't claim to have a complete answer to this question, but the concept-study approach that is described in the remaining chapters is one model that, we believe, takes into account current understandings of the constructs discussed in this chapter.

Learning

Contemporary cognitive scientists join with some of the last century's greatest students of learning – Piaget, Vygotsky, Bruner, and others – in the conviction that learning is perhaps the most complex and among the most poorly understood of "familiar" phenomena. As learners who exist within learning systems, one would think that humans would have a good grasp of what learning is all about. But the opposite seems to be the case.

One obstacle that must be addressed in efforts to understand the complexity is the deceptive simplicity of popular interpretations. In particular, English has three principal metaphors for learning: acquiring ("getting the idea," "taking things in," etc.), traveling ("getting there," "making progress"), and constructing ("building understanding," "assembling ideas"). These are so pervasive that, for the single-language speaker, it can difficult to conceive of learning in any other terms – even though, in learning, the learner actually acquires nothing, travels nowhere, and constructs nothing.

Within a complexity frame, the principal metaphor for learning is coherence-maintaining. That is, learning refers to any adaptive process through which a learning agent maintains its internal coherence (i.e., among the agents it comprises) and/or its external coherence (i.e., as part of its dynamic, evolving context). This notion applies to neurons within a neuronal cluster, to humans within social groupings, to species within ecosystems, and so on.

For the present purposes, our interest in teachers' disciplinary knowledge of mathematics prompts our attentions to three learning systems in particular: individuals, social groupings, and M_4T itself. We offer a bit more detail on this statement below.

Learning: analogical v logical

Mathematics curricula – and, for that matter, the curricula of most subject areas in the modern school – tend to be organized around the assumption that humans are logical creatures. Following the model of the rational-deductive proof, as exemplified in Euclid's geometry, the essential idea is that one starts with the most rudimentary parts of a system and builds in a logical, sequential manner into more sophisticated, defensible truths.

In terms of underlying metaphor, the form that is most often invoked in discussions of curriculum and learning is not the deductive proof but the image of building – which is part of the reason that there is such an easy fluency in educational discussions with such terms as basics, foundations, building blocks, construction, scaffolding, and upward progress. A secondary image for learning in popular discussions is that of a route between two points, by which learning is construed in terms of forward, incremental movement along a pre-specified path. (The word *curriculum* is derived from a Greek term meaning "path to be followed.")

These metaphors are so commonly invoked that it can be difficult to recognize that they are figurative and not literal descriptions of the structures of knowledge and the dynamics of learning. However, now that systems of knowledge can be analyzed for their networked structures and brains can be observed in real time, it's clear that there are no pristine edifices under construction and no straightforward paths to be pursued. Rather, knowledge production and learning seem to be more about establishing ever-more complex webs of connection – and, for the most part, these connections aren't logical at all.

Rather, they tend to be circumstantial, accidental, and at times almost random. The good news is that many of the associations that learners make are fairly consistent across particular cultural or social groupings, owing to entrenched habits of interpretation that are embedded in the language or pervasive in the physical space. Good examples of such *common* sense are the suggestions that knowledge is a building and that learning is construction. At the other extreme, as any experienced teacher will attest, learners often make entirely idiosyncratic associations among concepts and happenings despite similar experiences and near-identical vocabularies.

The important point at this juncture is not the truism that humans are predisposed to making connections among their experiences; nor is it that this predisposition is actually the principal mode of human learning. Rather, the critical point for now is that the main mechanism in abstract learning is figurative association – using metaphor, metonymy, image and similar strategies. That is, humans are mainly analogical, not logical (Lakoff & Johnson, 1999). In fact, while logical feats are easy for machines, we humans find them horrendously difficult in comparison to, say, the ease with which we perceive and project conceptual relationships among warm weather, passion, anger, contraband, and trendiness (which, for English speakers, cluster together around the notion of "hot").

Briefly, then, humans are capable of logic – but that capacity seems to ride atop the irrepressible tendency to make connections. Moreover, when the language of rationality and deductive argument is carefully deconstructed, "logic" itself proves to be highly analogical. As Lakoff and Johnson illustrate, conceptions of what it means to be logical, deductive, and rational are thoroughly anchored to figurative notions of containment and linearity.

To be sure, logical thought is much more than a simple figurative device. It is a powerful conceptual extension of a biological predisposition. It transcends inborn capacities, it doesn't piggy-back on them. Logical thought, then, should not be construed as some sort of secondary or subsidiary function. It is an extraordinary achievement.

However, when it comes to learning and teaching, the centuries-old habit of organizing schooling experience around the assumption of a core logical nature is highly troublesome. Somehow events to support mathematics learning must be structured in ways that recognize humans' penchant for analogical thought at the same time as they capitalize on humans' potential for logical thought.

In fairness to the educational literature, some major strides have been made in this direction. This point is reflected in Tasha's anecdote through her second attempt at teaching what it means to divide by zero. Consistent with a well-articulated sensibility, as noted in the narrative, she presented her students with some explicit background on varied meanings of and models for division. Yet for reasons that weren't clear to her at the time, students didn't seem to connect with these explicit references.

But they did the following year, we suspect in large part because Tasha had, by then, developed a different sort of expertise.

Learning: surface v deep

In 1976, Marton and Säljö offered educators a set of labels to differentiate between two very different ways of engaging with subject matters: deep and surface learning. In our experience, every experienced teacher resonates with this distinction. The same seems to be true of every student who hears about it.

Surface learning relies on rote. "Knowledge" is seen as "information" that is accepted passively, viewed uncritically, left disconnected from other information, and kept accessible only as long as it is needed to pass an exam or earn a credential. Deep learning revolves around looking for connections among ideas, finding the core argument, separating argument from evidence, hooking to other knowledge, and linking to life beyond the knowledge domain.

Not surprisingly, one's mode of engagement has much to do with one's purposes for learning. At times, surface learning is all that's needed – for instance, when figuring out how to use a new gadget. When the activity is ends-driven, or learning curves are too steep, or time is too limited, or anxieties are too high – the list goes on – one might expect a more instrumental, disengaged attitude and a more superficial learning.

For mathematics educators, however, surface learning is rarely appropriate – or at least the rhetoric around school mathematics paints a more hopeful picture. No doubt, most mathematics educators desire to instill curiosity in students and to help them to connect what they're doing in math class to previous knowledge, to other courses, and to life outside of school. We certainly do, yet at the same time we despair over the disconnects between such noble goals and the institutional structures that press students toward a more mechanical attitude.

That said, there are certainly practices and emphases within teaching that can profoundly influence students' mode of engagement. On this count, we find Dweck's (2006) research into the different mindsets of learners to be particularly useful, as it hints at aspects of teachers' disciplinary knowledge that are not often discussed. For instance, teachers can suppor "growth mindsets" and promote deep learning among their students as they pursue learner questions, engage students in debates, introduce the personalities associated with the developments of various concepts, foreground tensions and conflicts in the history of a subject area, and otherwise highlight the humanity of a discipline. By contrast, teachers who follow the text, prepare for the exams, treat the subject matter as ahistorical and abstract, and rarely veer from the well-timed trajectories of their lessons nurture more "fixed mindsets" and promote more surface learning.

This last paragraph, we believe, should highlight a fundamental tension around mathematics learning, impacting attitudes and approaches to both student and teacher knowledge. It all hovers around the word *humanity*, and the need for teachers to hold their disciplinary knowledge in a way that permits learners to engage *as humans* with concepts. Effectively, Dweck's research reflects an earlier warning from Davis and Hersh (1986):

> A detemporalized mathematics cannot tell us what mathematics is, why mathematics is true, why it is beautiful, how it comes to be, or why anybody should care a fig about it. But if one places mathematics squarely within human time and experience, it becomes a warm and rich source of possible meanings and action. Its ultimate mystery is never dispelled, yet it is exhibited as one of the prime creations of the human intellect. (p. 201)

Very little of the current research into teachers' disciplinary knowledge of mathematics actually touches on these human dimensions. As far we're aware, there isn't much classroom-based research on the matter either.

Nevertheless, we would assert that Tasha's third approach to teaching the concept of dividing by zero comes close to making mathematics "a warm and rich source of possible meanings and action." Her invitations to have students consider the images and metaphors that frame their understandings of division and zero enabled them to move beyond a memorized rule (surface learning) to an actual conceptual appreciation of what might be sitting behind that rule (deep learning).

Significantly, it was not just the conceptual emphases of the lesson that enabled this deepened learning. The interactive structures for learners, we suspect, were at least as important.

Learners: individual v collective

What's a learner? A few decades ago this question would've been heard as absurd in educational circles. "Learner" was synonymous with "student" or "pupil" – at least, ideally. These words were used interchangeably to refer to the individual who was supposed to be learning.

That's still the case in most popular discussions of schooling, but it's not how the word is being used in this text. Aligning ourselves with complexity research (e.g., Mitchell, 2009), we understand a learner in terms of self-regulating, self-determining, adaptive, emergent behavior – a definition that, in turn, opens up the word *learner* to any coherence-maintaining, self-changing system.

In our experience, this redefinition of "learner" would be seen by most as one of the more radical suggestions in this text, but it's not an altogether new idea. Its history has been traced in some detail elsewhere (e.g., Davis & Sumara, 2006; Davis, Sumara, & Luce-Kapler, 2008), and it is important to note that it is a notion that has been creeping quietly, but pervasively into the mathematics education literature. For example, it is not uncommon to encounter suggestions that mathematics itself "is a living, breathing, changing organism ..." (Burger & Starbird, 2005, p. xi) or that it "emerges as an *autopoietic* [i.e., self-creating and self-maintaining] system" (Sfard, 2008, p. 129). That is, as we develop in more detail at the start of the next chapter, the actual discipline of mathematics might be productively construed as a learner.

On a perhaps more readily understood level, the classroom collective might also be construed in such terms – that is, not merely as a collection of learners, but as a collective learner. This point is demonstrated every day in school staff rooms, as teachers make reference to groups of students as single agents (e.g., "7A works really hard," "9B is easily distracted," "8C is the brightest class I've ever had"). Indeed, Tasha did just this in her comment that "the class could not explain why" division by zero couldn't be done. On that note, it is interesting to compare the interactive structures of Tasha's three lessons. Something decidedly different was going on in the third one, in which students were working together to create a group-based interpretation that surpassed what any individual could offer on her or his own. That is, in contrast to the first two attempts at the topic, in the third iteration of the lesson the conditions were in place for the group to be smarter than the smartest person in the group.

This point is a vital one. With the strong undercurrent of individuality in western sensibilities, the suggestion that attention be diverted to the group is often heard as an attempt to diminish the uniqueness of the person. Exactly the opposite is intended here. In fact, it is precisely the uniqueness of individual contributions that define the limits of collective intelligence. A group in which everyone contributes the same thing will be no smarter than the typical contribution. A group that offers and grapples with diversity of interpretation has potential to go to interesting new places. In other words, "the best way for a group to be smart is for each person to act as independently as possible" (Surowiecki, 2004, p. xx) – where "independence" is in reference to personal choice, not individual isolation. Group-based intelligence is not rooted in a this-or-that logic. The possibilities for the individual learner and the collective learner can and should amplify one another.

This particular topic has been discussed in a number of ways in the educational literature. Perhaps the most popular frame at the moment is *participatory epistemologies* or *participatory cultures*. Within these discussions, as

Jenkins and colleagues (2006) highlight, there are emphases on the co-creation of knowledge, creativity, and innovation – all with a dimension of critical thought that allows for the development of deep, conceptual and relational understanding. To that end, and looking forward to the next few chapters, we propose that M_4T is a learner, and so it demands collective, participatory engagements. The structures we propose to support the emergence of this learner are thus guided by these assumptions:

- Individual and collective knowing cannot be dichotomized; collective possibilities are enfolded in and unfold individual understandings.
- M_4T is too vast and too volatile to be considered in terms of mastery by any individual. Rather, it is simultaneously an individual and collective phenomenon.
- At the individual level, understandings of mathematical concepts and conceptions of mathematics are emergent.
- At the level of social collectives, teachers' knowledge of mathematics is largely tacit but critical elements of it can be made available to conscious interrogation in group settings.
- At the cultural level, teachers are vital participants in the creation of mathematics, principally through the selection of and preferential emphasis given to particular interpretations over others.

To situate these remarks, we re-emphasize that much of formal education places the individual learner in the centre of the educational process. But humans learn collectively by participating in activities of their cultures/ societies. We feel it is important to place much more emphasis on the collective dynamics of individuals – that is on collective learning – while preserving the obligation to help individuals develop competence. To that end, in the chapters that follow, we offer one model for research into and development of teachers' disciplinary knowledge of mathematics that allows participants to specialize and, through that specialization, contribute to collective potential.

Where Are We Going with This?

This chapter is intended as much more than a list of developments that we see as game-changers within education. Rather, it is intended more as a necessary link between Chapter 1 (our account of the evolution of thinking around teachers' disciplinary knowledge of mathematics) and the remainder of the book.

To underscore some key elements, perhaps the most important game-changer in recent discussions of knowing and learning is one that we've left implicit through the chapter – namely that what is called "knowledge" is usually perceived as stable, whereas what is described in terms of "knowing" is seen as dynamic. We regard this distinction as arbitrary, an artifact of human temporal frames.

This is not in any way to deny the stability of knowledge. Rather, the point is to highlight the importance of recognizing its evolving character. Vis-à-vis mathematics, it's useful to know how new concepts emerged and how old ones were elaborated to accommodate emergent needs, situations, and possibilities … as such *knowledge* enables us to think about how individual *knowing* arises.

We aim to illustrate precisely this point in the chapters that follow, at the same time as we present our case for the suggestions that both the discipline of mathematics and collectives of individuals can be seen as coherent learners. This point, in particular, is pivotal to understanding our own take on the question of teachers' disciplinary knowledge of mathematics.

So, what is our take on the M_4T question?

To recap, we see research over the past 40 years to have been organized around an evolution through the following:

Q1) What *mathematics* do teachers need to know in order to teach mathematics? (1970–1990-ish)

Q2) What *specialized mathematics* (i.e., PCK) do teachers need to know in order to teach mathematics? (1990–2000-ish)

Q3) What mathematical knowledge *is entailed by the work* of teaching mathematics?" (2000–current)

We embrace all of these questions as necessary and important foci in the field's collective research in teachers' disciplinary knowledge of mathematics. To them, we would like to add our own emphasis:

Q4) How must teachers know mathematics for it to be activated in the moment and in the service of teaching?

SUBSTRUCTING EMERGENT MATHEMATICS

CULTIVATING AN OPEN DISPOSITION

In brief …

This chapter introduces *concept study* – a participatory methodology through which teachers interrogate and elaborate their mathematics. We use a particular concept-study event to highlight two important characteristics of all concept studies: the process of substructing mathematical knowledge and the emergent nature of mathematics. In the process, we further develop the suggestion that M_4T is better construed as an open disposition towards mathematics in educational settings than a specifiable collection of skills or competencies.

Is 1 Prime?

[VOICE OF BRENT[1]]

In the spring of 2007, during a concept study that involved eleven middle school teachers, one of the teachers raised the question, "Is 1 prime?" The response proved to be a little surprising.

At the time the question was raised, the group had already been meeting for about four hours each month for over a year. Prior to the discussion of the primeness of 1, the participants had explored the question, "What is multiplication?" at some length.

Over a number of meetings, the group members had assembled the following set of interpretations of multiplication:

- *repeated addition;*
- *grouping;*
- *number-line hopping;*
- *number-line stretching or compressing;*
- *branching;*

---

1. See Davis (2008) for a more complete and detailed account of this event.

- *making grids or arrays;*
- *area-making, volume-making, and dimension-jumping;*
- *steady rise (slope);*
- *proportional reasoning; and*
- *number-line rotation (for integer multiplication)*

This list served as the starting place for extensive discussions of the figurative bases of mathematical understanding and the varied entailments of different metaphors and images. But the discussions didn't really heat up until one of the teachers expressed a concern about his inability to offer an adequate explanation to his 6th-grade students of why 1 isn't considered a prime number. Of course, every teacher in the group was aware that all standard textbooks maintain that 1 is not a prime number and a few were aware that the Fundamental Theorem of Arithmetic excluded 1 as a prime.[2] But why is this so?

The group members were eager to engage with the question, and their prior study of multiplication proved very useful in the session that ensued. They decided to examine the mathematical notion of primeness through the diverse metaphors of multiplication. Figure 3.1 shows a portion of the "entailments chart" that the teachers created to aid their thinking.

The complete entailments chart created by the teachers was several times larger than the portion presented here, and covered a dozen interpretations of multiplication. However, the two shown in the figure are sufficient to highlight the issue that was uncovered: different metaphors and interpretations of multiplication lead to different answers to the question, "Is 1 prime?" In fact, the metaphor of rectangular arrays, which is regularly used by many teachers in discussions of primes and composites, can actually support to the "undesirable" conclusion that 1 is prime.

It was interesting to note that none of the teachers was perturbed by the conflicting results. One participant put it this way: "No wonder that question [of 1 being prime] always bothered me!"

The concept study of primeness continued for a couple more meetings, during which the group compared several dictionary definitions of "prime." These studies revealed that, prior to the 19th century, most mathematicians considered 1 to be prime (Gowers, 2002). By the end of the inquiry, the consensus among group members was that students should also be involved in some manner of concept study, as doing so would highlight that the (historical) development of mathematics involves debate, analogical reasoning, and conscious selection.

---

2. The Fundamental Theorem of Arithmetic (which is also known as the "unique-prime-factorization theorem") states that any integer greater than 1 can be written as a *unique* product of prime numbers. That is, this theorem excludes 1 from the set of primes. If 1 were not excluded, every integer greater than 1 would have multiple prime factorizations. For example, 6 might be factored as 2×3, as $1 \times 2 \times 3$, as $1 \times 1 \times 2 \times 3$, etc.

If multiplication is ...	... then a factor is ...	... then a product is ...	... then a prime number is ...	... then a composite number is ...	Is 1 prime?
DOING A SEQUENCE OF FOLDS	an n-fold	the count of regions bounded by creases from folding	a product that can only be reached by folding directly	a product that requires a combination of folds	No. Since no folds are involved in making a product of 1, it is neither prime nor composite.
MAKING A RECTANGULAR ARRAY	a dimension	the count of objects/cells in the array	a product that will always have a dimension of 1, no matter how you arrange it	a product that can be arranged in an array in which neither dimension is 1	Yes. A product of 1 must have a dimension that is 1.

Figure 3.1. A portion of an "entailments chart" used to inquire into "Is 1 prime?"

How must teachers know mathematics for it to be activated in the moment and in the service of teaching?

To begin to answer this question, it is important to be attentive to some of the subtle complexities of teachers' disciplinary knowledge of mathematics. For us, the above account of the primeness of 1 gestures in this direction. While the need for teachers to possess specialized knowledge of formal mathematics is indisputable, the range of interpretations of concepts that might be at play in any given pedagogical moment is simply too broad, too varied, and too context-dependent to be pre-specified and mastered. The complexity of teachers' disciplinary knowledge demands an open disposition toward the mathematical interpretations and meanings present, whether they are

- rooted in pre-established formal mathematics;
- selected and adapted by teachers to make mathematical concepts more accessible;
- invented by students in efforts to render mathematics personally coherent.

We have separated these "sources" not because we think they're non-overlapping but because we feel it's important to acknowledge the range of origins of mathematics meanings in the classroom. We see such diversity as a resource to develop rather than a problem to manage, following one of the major conclusions of recent cognitive science research. It turns out that conceptual diversity is the norm, not the exception (see, e.g., Gibbs & Colston, 2007). In other words, mathematical sense making is not about the *best* image, the *proper* interpretation, or the *correct* metaphor. It is about the nurturing of *adequate*, *appropriate*, and *useful* interpretations for the tasks at hand.

Closer to the ground of the classroom, perhaps, this emergent insight might be taken to suggest that the learning of mathematics should be more structured around *meanings* than *definitions*. That is, few would attempt to understand "plate" or "barrier" by defining it and then seeking out instances of the definition. Rather, such notions arise organically through specific meanings gleaned from encounters in the world. Usually these meanings eventually cohere into a notion that might be defined – but things don't often unfold in the other direction (i.e., from formal definition to rich meaning).

As the above account of the primeness of 1 illustrates, the methodology of concept study is simultaneously about meanings and definitions. More descriptively, it is a structure that is intended to provide teachers with the sorts of experiences and attitudes that might cultivate disciplinary knowledge founded on conceptual diversity. We will now move on to take a closer look at this methodology.

Concept Study

The phrase "concept study" combines elements of two prominent notions in contemporary mathematics education research: *concept* analysis and lesson *study*. Concept analysis, which was well represented in mathematics education research from the 1960s to the 1980s, focuses on explicating logical structures and associations that inhere in mathematical concepts (e.g., Leinhardt, Putnam, & Hattrup, 1992). As Usiskin and colleagues (2003) described it, concept analysis

> involves tracing the origins and applications of a concept, looking at the different ways in which it appears both within and outside mathematics, and examining the various representations and definitions used to describe it and their consequences. (p. 1)

Usiskin et al. extended their description to include ways of representing ideas to learners, alternative definitions and their implications, histories and evolutions of concepts, applications, and learners' interpretations of what they are learning. Across the past half-century, concept analysis has been untertaken principally by university-based experts, with insights disseminated through academic journals and course textbooks. In constrast, the researchers in concept studies are practicing teachers.

That is, in concept study, we blend the foci of *concept analysis* with the collaborative structures of *lesson study*, a well-known collaborative structure that "teachers engage in to improve the quality of their teaching and enrich students' learning experiences" (Fernandez & Yoshida, 2004, p. 2). Lesson studies are oriented towards new pedagogical possibilities through participatory, collective, and ongoing engagements. Sharing the emphases on collective process, concept studies are more focused on the actual mathematical content of teaching. They are intended to prompt teachers to go beyond well-rehearsed elements that are already structured into mathematical concepts as part of the culture of school mathematics. They are, in effect, moments of collective didactical transformation – opportunities to work together to re-form concepts in ways that render those concepts more accessible to learners.

The idea of concept studies emerged rather unexpectedly from research work in the early 2000s when some funding made it possible to meet regularly with groups of practicing teachers to discuss trends in mathematics education, to engage in problem solving, and to examine topics that participants found challenging (see Davis & Simmt, 2006). During this 3-year project, conceptual mathematical issues (e.g., metaphors of multiplication, division by 0) kept coming to the fore. They challenged the participants (including the researchers) to identify effective collective structures for engaging with them.

Typical meetings involved 20 to 30 teachers who were drawn from all grade levels. The fact that teachers from Kindergarten through high school were taking part was actually accidental. The school district in which the first studies were conducted was quite small, and in order to meet the participant quota specified in the funding, it was necessary to extend the invitation to any teacher involved in mathematics. This unintended "accident" afforded one of the most important features of concept study, as the cross-grade expertise represented in the collective turned out to be a vital factor in the emergence of shared appreciation of the conceptual breadth complexity of school mathematics. In particular, three consistent consequences of the cross-grade representation emerged:

- As a result of working together, participants were able to demonstrate a detailed knowledge of the K–12 experience of students and the structured progression of mathematical concepts through the grades. Such *vertical* (i.e., across-grade) *awareness* arose as primary teachers discussed mathematical concepts with high school teachers, learning from one other what students will have experienced in earlier grades and what they will be expected to know in later ones.

- Opportunities to discuss concepts among colleagues who taught mathematics at different levels afforded the development of *horizontal* (i.e., within-grade) *interpretive breadth*, as it prompted participants to renew awarenesses of habits of defining and illustrating. As they listened to contributions from colleagues who taught other grades, teachers realized that they often invoked multiple interpretations without conscious awareness. This point is illustrated by the range of metaphors of multiplication in this chapter's opening account. As we will show later, such metaphors often cluster in very coherent (but rarely explicit) ways at different grade levels.

- As a collective, participants began to make conscious and deliberate efforts to frame understandings and explanations of concepts in terms of *open definitions* – that is, in ways that were adequate to the grade levels they were teaching, but that also anticipated the possibility of future elaboration. Multiplication is a good example of this practice. In the early meetings, elementary school teachers usually stuck to the metaphor of "repeated addition," as they believed that it provided a comprehensive definition for the concept of multiplication. But as the sessions progressed, and these teachers were confronted with questions of how one might multiply fractions, signed integers, irrational numbers, and algebraic terms, they started to take greater care and to think through more nuanced interpretations of multiplication.

To be clear, these three consequences – vertical awareness, horizontal interpretive breadth, and open definitions – were neither anticipated by us nor were they deliberate. They were emergent consequences of the diversity represented in our early concept study groups. More recently, as will be elaborated in Chapter 4, we have incorporated such elements more deliberately into our methodology as we engage in concept studies with new groups of teachers.

Early concept study groups presented us with some noticeable organizational challenges and frustrations. Most problematic was the issue of irregular participation. It was typical for the composition of the group to vary by as much as 50% from one meeting to the next. The lack of continuity made it nearly impossible to build up shared understandings[3] among the group members. The participants' degree of commitment to concept study also varied. Some sought out resources outside the meeting times, studied in advance, and adapted strategies for use in their classrooms. Others just showed up to the concept study meetings. As a result of these inconsistencies, it was difficult for us to track the influence of our early concept study work on the practice of the teachers.

The concept study group involved in the "Is 1 prime?" exploration was a transitional group, with which we sought to address some of these earlier frustrations. Its meetings were mandated as part of a school-board certification program that required practicing teachers to advance their academic qualifications, and so the eleven teachers attended regularly and expected to do "out-of-class" work. This structure afforded a continuity and emergence of shared understandings that enabled the type of examination reported in this chapter's opening narrative. Unfortunately, the certification program included only middle school teachers. The reduced diversity contributed to a noticeable loss of vertical awareness among the participants (in comparison to the insights demonstrated by participants in earlier K–12 groupings). As well, because the concept studies were part of a certification

---

3. "Shared understanding" and related terms have been the topic of considerable discussion among mathematics education researchers. In particular, the suggestion that meanings and interpretations might be shared among different individuals has been criticized my many as naïve (see, e.g., Cobb, Wood, Yackel, & McNeal, 1992).

We concur, insofar as "shared" is taken to mean "identical." Clearly, if meanings are assembled from experience, it would be impossible for different individuals to assemble identical senses. However, if "shared" is taken in its more common sense of "having a portion of something with another or others," the notion of shared understanding takes on a very different meaning – and one that is more resonant with complexity thinking. We use the term in the latter sense here, asserting that understandings and meanings are shared/distributed across communities. Understandings are shared in the ways that meals, stories, rides, and emotions are shared.

program and not a research program, we were unable to examine the impact of the work on teaching practices.

Some more recent iterations of concept study groups have been organized as a graduate certificate and degree programs at the University of British Columbia and the University of Calgary. We will report on this more deliberate approach to concept study in Chapter 4. For now, we note that its key elements are, to a large extent, formalizations of structures, emphases, and strategies that were developed with teachers in earlier concept study groups. To put it differently, the structures of concept study emerged from actual work with teachers, and continue to evolve and be refined as we work with more groups of teachers.

In anticipation of the next chapter's discussion of the structures of formalized concept studies, we devote the balance of this chapter to three essential qualities of concept study witnessed across all the groups with which we worked: substructing, emergence, and open dispositions.

Substructing

The word *substructing* was first used in the context of a concept study by the same teacher who posed the "Is 1 prime?" question. He wasn't actually sure it was a word when he first used it, but he felt the need for a term to contrast what the group was doing to act of *unpacking*.

"Unpacking" mathematical constructs is one of the more popular notions to have arisen in recent discussions of teachers' disciplinary knowledge of mathematics (e.g., Ball, Hill, & Bass, 2005; Hill, Ball, & Schilling, 2008; Ma, 1999), and it is used to describe an aspect of mathematics knowing that is unique to teachers. Whereas an important component of research mathematicians' work is to collect their thinking into compact formulations – that is, packing – it is teachers' task to perform the reverse operation. Teachers must be able to take apart formulae, operations, and mathematical terms, so that students can gain access to the thought processes and ideas that they represent.

Concept studies undoubtedly include some unpacking activities. As illustrated in the opening narratives of Chapters 2 and 3, when seeking to make sense of a complex mathematical idea, one useful starting point is to generate lists of metaphors, analogies, and images that might be associated with that idea. The process of generating such lists both renders explicit the primarily tacit nature of human knowing and the principally analogical nature of human learning. The main aim of unpacking activities within concept studies is to recall the figurative aspects of understanding, which expert knowers might have forgotten they know.

However, what follows in concept studies cannot be aptly construed in terms of unpacking. As the "Is 1 prime?" example shows, mathematical concepts transcend the elements included in previously packed representations. Deep understanding of a concept requires more than pulling apart its constituent parts; it also entails examinations of how these parts hold together and fall apart in different contexts and circumstances. Teachers' engagement in such examinations often leads to generation of novel insights that transcend pre-established and recovered packed insights. As the "prime" example illustrates, the process can lead to a reworking of the concept itself. That is, at the end of a concept study, the concept under investigation is transformed for the teachers who engaged in its inquiry. And this is precisely the sense intended by the teacher who offered the term; for him, substructing suggested a sense of dismantling and rebuilding. For him (and, very quickly, for the group), it highlighted the creative dimensions that inhere in the reworking process and distinguish them from the merely descriptive/interpretive emphases of unpacking.

It turned out that his meaning of dismantling and rebuilding is very close to the dictionary definition of the word. *Substructing* is derived from the Latin *sub-*, "under, from below" and *struere*, "pile, assemble" (and the root of *strew* and *construe*, in addition to *structure* and *construct*). To substruct is to build beneath something. In industry, *substruct* refers to reconstructing a building without demolishing it – and, ideally, without interrupting its use. Likewise, in our concept studies, teachers rework mathematical concepts, sometimes radically, while using them almost without interruption in their teaching.

While the term "substructing" may connote a sense of downward direction or motion, the process is both reductive and productive. It is reductive in that it starts by re-collecting and re-membering experiential, linguistic, and other elements – that is, the *meanings* – that infuse an understanding of a concept. (For example, such reduction was demonstrated by the "Is 1 prime?" group through their list of interpretations of multiplication.) It is productive in the sense that such acts of re-presentation often compel new integrative structures and novel interpretations. (For example, such production was demonstrated when the group looked across entailments of different interpretations to select meanings appropriate to standard definitions of *prime*.) These constructs, in turn, may become substructs of subsequent knowing. This recursive process corresponds with the understanding of emergent knowledge as both dynamic and stable/coherent: always deepening, crystallizing, and becoming, while always embodying the memory of its evolution in its structures.

Another way to understand the differences between unpacking and substructing is to think about them in terms of Kahneman's System 1 and

System 2, discussed in Chapter 2. In our experience, unpacking activities during concept studies are most closely associated with System 1 – the fast, automatic, memory-based part of one's knowing. Unpacking is about digging out what is already there, albeit "buried" or "obscured." In contrast, substructing mainly involves System 2 – the slow, reflective, thinking-based part of knowing. Relatively little of the work of substructing is recollection. Rather, it is more about re-interpreting, reconfiguring, reknowing what is assumed to be known already.

When construed in this way, the process of substructing operationalizes a phenomenon that is always at work within systems of mathematics, collective sense-making, and personal learning – namely, emergence.

Emergent Mathematics

The notion of *emergence* has been implicit in just about everything that we've discussed so far. The term was co-opted in the 1980s by researchers attempting to describe the dynamics of self-organizing and self-maintaining systems, such as biological cells, consciousness, and economies. With such phenomena, the definition of emergence is clear: emergence is what has happened when a coherence (i.e., a new entity, such as a an economy, a thought, or an ecosystem) has arisen in the activities of interacting agents.

Many popular and highly accessible accounts of emergence have been published over the past few decades (e.g., Johnson, 2001; Mitchell, 2009; Waldrop, 1992), and we recommend these introductory texts to readers who are unfamiliar with the history and foci of the complexity sciences. Studies of emergence are oriented by the assumptions that that there is a fundamental unity to life and that different living systems exhibit similar patterns of organization (Bejan & Zane, 2012; Capra, 2002). These assumptions are supported by the major findings of complexity science, systems theory, and cognitive sciences of the past three decades. Accordingly, the pattern of organization of living systems is the self-generating network. Living systems are cognitive learning systems, where cognition is closely related to the process of *autopoiesis* (self-organization, self-generation) and can be observed as *evolution* on multiple scales.

Complex living/learning systems are coupled to their environments and continually respond to environmental influences with structural changes. They are dissipative structures that operate far from equilibrium where new forms of order may emerge. As Goldstein (1999) explained, emergence is "the arising of novel and coherent structures, patterns and properties during the process of self-organization in complex systems" (p. 51).

In terms of metaphor – and, in particular, visual metaphor – two images have risen to particular prominence in efforts to depict complex emergence. We have been making extensive use of one, namely "embedded circles," to illustrate the relationships among already-emergent systems. Another image that has is used more to illustrate the actual process of emergence is that of branches branching into branches – that is, the tree. It turns out that efforts to map out histories of cultures, evolutions of species, genealogies of ideas, and so on all seem to gravitate to the same sorts of bifurcating images. Moreover, nature shares this penchant for branching forms, which appear wherever energy or matter dissipate into grander spaces (see Bejan and Zane, 2012). While not a hallmark, then, a tree-like pattern is certainly a clue that something emergent might be happening as it signals expanding possibility and mounting complexity.

The ubiquitous tree images in complexity research also signal a deep connectivity in complex, emergent forms. Describing a phenomenon as emergent is tantamount to saying that it manifests properties that do not exist among its separated components – that, in effect, a tree is more than a bunch of connected sticks. We believe that the notion of emergence can be applied effectively to a description of mathematics for teaching. Conceiving of M_4T in as an emergent phenomenon highlights not only situation-specific selections and adaptations, but also the systemic, distributed, and self-organizing character of teachers' content knowledge.

In an early draft of this chapter, we attempted to illustrate this point by mapping the "Is 1 prime?" activity onto a tree diagram, which we then intended to lay into a tree-based image of arithmetic. The task quickly overwhelmed us, and the reasons for our lack of success are instructive. The problem wasn't that a tree image was inappropriate for tracing the group's thinking. Rather, the issue we confronted was the incredible diversity of interpretation that turned out to be knitted into something as simple as multiplication. Our generated image was very quickly meaningless by virtue of its branching complexity. Trees of knowledge seem to be more complex than oaks and poplars.

Even so, for us, the image of growing and constantly bifurcating branches of thought is implicit in every aspect of concept study. In particular, we use the notion of emergence in our concept study research to understand and characterize different co-implicated systems within M_4T:

- the evolving structures of teachers' personal understandings;
- the knowledge-producing dynamics within social groupings of teachers; and
- the dynamics and structures of a domain of knowledge – namely, mathematics.

There is nothing particularly provocative in the suggestion that personal understanding and social co-activity are dynamic and adaptive phenomena. These ideas have become commonplace in the mathematics education literature, primarily through radical constructivist (e.g., von Glasersfeld, 1990) and social constructivist (e.g., Cobb, 1994) discourses. What might be somewhat more contentious is the suggestion that all domains of knowledge – including mathematics – fall into the category of emergent phenomena. Whether mathematics is discovered or created has been the topic of intense philosophical debate for centuries, and we don't propose to resolve the matter definitively with the suggestion that mathematics is emergent. However, we believe that it's important to acknowledge that collective human understanding of mathematics is an evolving complex form (see Foote, 2007), and that it is precisely this understanding that is of primary concern to educators.

As an example of the evolving nature of human conceptions of mathematics, consider Mazur's (2003) historical case studies of the evolution of mathematical concepts. Multiplication, for instance, was likely originally conceived as repeated addition and/or some sort of a grouping process, and was applied to the set of counting numbers. However, these metaphors begin to break down when one attempts to multiply fractions by repeatedly adding fractions a fractional number of times. They become virtually useless when one multiplies irrational numbers or imaginary numbers.

Mazur recounts mathematicians' struggles to elaborate multiplication as new number systems unfolded. Their responses were never purely logical and, in most cases, new realizations – that is, associations that might be used to interpret a concept, including definitions, rules, analogies, applications, and so on – emerged and were incorporated into always-evolving conceptual constructs. For instance, construing multiplication as area making, wherein the factors are dimension (lengths), makes it is easier to interpret and envision calculations that involve fractions and irrational numbers. Other realizations are needed to make sense of multiplication of negative integers and imaginary numbers. In effect, multiplication, and indeed all mathematical concepts, have "open definitions." Put differently, a mathematical concept must be defined in a manner that is sufficient for the situation at hand, but that permits elaborations as new situations present themselves.

It bears mention here that the practice of weaving knowledge of such historical details into mathematical instruction has been associated with increased student engagement and efficacy (Dweck, 2006; Marton & Säljö, 1976a, 1976b). We suspect that this is due to the relaxation of the monological stranglehold of formal mathematics. Regardless of what one believes about the ontological status of mathematics, the simple fact is that human interpretations of mathematics have been the sites of struggle – personal struggles to find sensible interpretations, interpersonal struggles to champion notations, and even cultural struggles to assert interpretive dominance. As we already mentioned in Chapter 1, mathematics' pervasive role in the new economy as a means to resources and power is undeniable. To omit or ignore such details is to deny the fundamental humanity of the enterprise.

Open Dispositions

We now return to the over-arching theme of this chapter – the suggestion that it is productive to construe teachers' disciplinary knowledge as an open disposition, and not merely as a mastery of a discernible body of knowledge.

We use the notion of open disposition to link our M_4T question ...

Q4) How must teachers know mathematics for it to be activated in the moment and in the service of teaching?

... to a modest revision of the preliminary answer we offered in Chapter 1...

A4) M_4T is an *open disposition* toward mathematics knowledge that enables a teacher to structure learning situations, interpret student actions mindfully, and respond flexibly, in ways that enable learners to extend understandings and expand the range of their interpretive possibilities through access to powerful connections and appropriate practice

To contextualize this discussion, we re-emphasize that efforts over the past several decades to describe and identify teachers' disciplinary knowledge of mathematics have overwhelmingly focused on individual teachers' explicit knowledge – in particular, on formal (course-based) content knowledge, specialized pedagogical content knowledge, and content knowledge entailed in the work of teaching. As we emphasized in Chapter 1, we regard all of these elements as vital aspects of M_4T. At the same time, we see them as inadequate characterizations. In our analysis, they fall short on two key elements, both pertaining to the complexity of teachers' disciplinary knowledge of mathematics.

First, we are troubled by the implicit assumption that the individual teacher is the locus of these four extensive bodies of knowledge. Mathematics is so vast that no one would ever imagine an individual could "know" the discipline in any comprehensive manner. Rather, we contend that mathematical knowledge must be understood as distributed across an extensive network of knowers. We're guessing that there is widespread agreement among mathematicians and mathematics educators on this point. However, when the topic shifts from *knowledge of mathematics* to *teachers' knowledge of mathematics*, few people acknowledge the possibility that this domain of knowledge is also vast, evolving, and distributed. Yet these are qualities that showed up in great abundance in every concept study we have witnessed.

Second, none of the aforementioned explicit categories or questions seems to capture the complex, emergent quality of teachers' knowledge – as it was enacted collectively, for example, in the account of "Is 1 prime?" True, one might read into the above descriptions the possibility that this knowledge evolves. However, we regard the mere suggestion of that possibility as patently inadequate. Rather, we argue, the complex, emergent character of teachers' knowledge is a defining quality of M_4T.

Phrased differently, we believe that lists of "types of knowledge" must pay heed to how teachers *are* with mathematics. Just as teachers need to frame mathematical ideas in terms of *open definitions*, they also need to embody *open dispositions* toward the subject matter and enact an openness to emergent possibility in their classes. As will be elaborated in Chapter 4, this open disposition entails a particular capacity to participate in a knowledge-building community through, for example, acknowledging and responding to potentially useful and possibly new mathematical ideas from students, using misconceptions as sites of learning, and inventing other strategies to extend the discipline beyond just formal mathematics.

It is important to note that an open disposition is very much dependent on formal (course-based) content knowledge, specialized pedagogical content knowledge, and content knowledge entailed in the work of teaching. However, it is not reducible to any of them. We believe that characterizing M_4T as an open disposition towards mathematics in educational settings provides educators and researchers alike with a broad enough scope for enacting emergent evolutionary possibilities in mathematics pedagogy.

Where Are We Going with This?

Concept studies began rather informally. Although they shared some common themes and activities – in particular, the search for tacit interpretations of different concepts – for most of the first decade of working with teachers, we did not have systematic strategies for substructing concepts.

That changed a few years ago. In the next chapter we begin to report on a few projects in which we distilled emphases and strategies, developed by and with teachers in earlier concept studies, to structure a more deliberate and systematic concept study.

CONCEPT STUDY

TEACHERS CO-CONSTRUCTING MATHEMATICS

— In brief …

This chapter recounts the movement to a more "deliberate" form of concept study, based on the emphases that emerged in less-structured settings. We offer four preliminary strategies: identifying extant meanings ("Realizations") analyzing the flow of these meanings within the curriculum ("Landscapes"), exploring their implications for applications and other concepts ("Entailments"), and blending them into more powerful constructs ("Blends").

Pulling Together, Not Just Pulling Apart

[VOICE OF BRENT]

The group of 11 middle school teachers that raised "Is 1 prime?" (see the opening anecdote of Chapter 3) had a talent for generating new strategies for grappling with vexing conceptual issues. One of those strategies was the creation of an "entailments chart" to analyze some of the implications of different interpretations of multiplication for the concept of prime. *Another emerged as they sought to better understand the relationships among meanings, rather than focusing so heavily on dismantling concepts.*

Recapping the last chapter's opening narrative, in an early meeting the group had generated the list of different interpretations of multiplication:

- *repeated addition;*
- *grouping;*
- *number-line hopping;*
- *number-line stretching or compressing;*
- *branching;*
- *making grids or arrays;*
- *area-making, volume-making, and dimension-jumping;*
- *steady rise (slope);*

- *proportional reasoning; and*
- *number-line rotation (for integer multiplication).*

Several months later, after the group participants generated similar lists for a handful of other mathematical concepts (including division, number, and equality), one of the teachers expressed a frustration: "I've really been appreciating the opportunities to dig into these ideas, but I'm wondering where I should be going with them. Should I be telling my students about all the bits and pieces and hope they put them together? Should I be putting them together for them? I'm not even sure that I'm putting them all together for myself. Maybe that's what I'm most worried about."

Not everyone shared this teacher's concern, but the participants agreed that everyone could benefit from efforts toward consolidating the many interpretations that had been generated. As one teacher put it, "we'll be pulling together, and not just pulling apart." After some discussion, it was agreed that, as a homework assignment, the group members would attempt to map out how the different interpretations of multiplication on the list might be related to one another.

A few weeks later, the group met again to share a collection of some very different mappings. The teachers had used a range of strategies to organize the interpretations. These included sorting them into grade levels, linking them to common applications, and classifying them as discrete/continuous and as enactive/iconic/symbolic.

In the discussion that ensued, the participants recalled that the point of the homework was "pulling together, and not just pulling apart." Yet, when viewed through the lenses of the new mappings, multiplication appeared more complex than ever. In order to counter the added complexity, the teachers decided to select only a few "critical dimensions," and to use them to construct one collective mapping. There was ready agreement that sorting the interpretations into the grade levels in which they are introduced was a good idea. The participants all thought that it would be useful to notice at what points in the K–12 sequence multiplication is elaborated significantly. After considerable debate on what should be the second organizing criterion, the group settled on sorting the interpretations into three types:

- arithmetic *(i.e., defined in terms of mathematical actions or objects);*
- partitional *(i.e., based on actions of separating); or*
- compositional *(i.e., based on actions of assembling).*

Using these categories as a horizontal axis and grade level as the vertical, after a few hours the group work created the "landscape" presented in Figure 4.1.

Upon examining their collective mapping, participants were surprised to notice that the different interpretations of multiplication, far from being random or isolated, were organized into grander interpretive structures. The new landscape

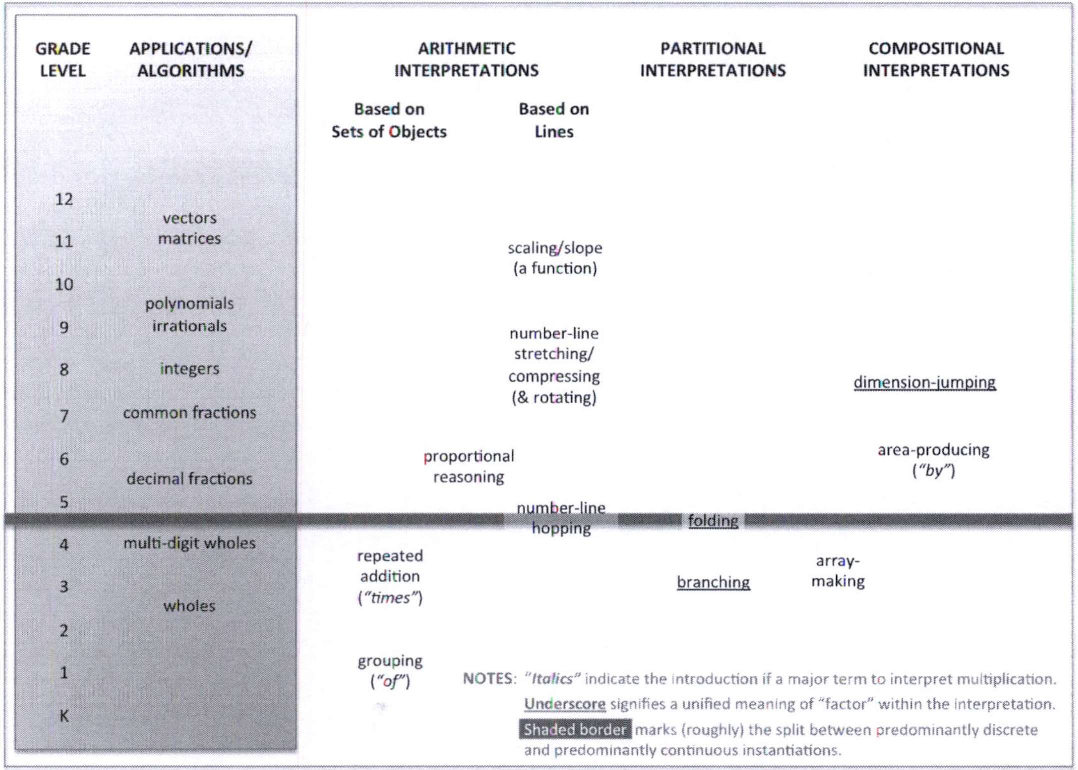

GRADE LEVEL	APPLICATIONS/ ALGORITHMS	ARITHMETIC INTERPRETATIONS		PARTITIONAL INTERPRETATIONS	COMPOSITIONAL INTERPRETATIONS
		Based on Sets of Objects	Based on Lines		
12					
11	vectors matrices				
10			scaling/slope (a function)		
9	polynomials irrationals				
8	integers		number-line stretching/ compressing (& rotating)		dimension-jumping
7	common fractions				
6	decimal fractions		proportional reasoning		area-producing ("by")
5			number-line hopping	folding	
4	multi-digit wholes			branching	array-making
3	wholes	repeated addition ("times")			
2					
1		grouping ("of")			
K					

NOTES: "*Italics*" indicate the introduction if a major term to interpret multiplication. Underscore signifies a unified meaning of "factor" within the interpretation. Shaded border marks (roughly) the split between predominantly discrete and predominantly continuous instantiations.

Figure 4.1. A Landscape of Multiplication

indicated that distinct and coherent strands of interpretation are systematically developed over the K–12 sequence. The teachers acknowledged that, through their teaching, they had been participating in this systematic development of multiplication, albeit without prior awareness of each participant's contributions in the grander scheme.

The structures of concept study have a ten-year history, and arose initially out of somewhat unstructured group meetings of teachers who shared an interest in understanding mathematics better (*cf.* Davis & Simmt, 2003, 2006). In this chapter, we describe these structures in their current format, that is, in the ways they are employed in recent and current concept studies. The opening narrative was chosen in order to emphasize that these structures and strategies were neither designed nor planned. Rather, they emerged spontaneously from the collective work of teachers in their efforts to come to grips with the meanings of mathematical concepts.

In 2009, we transitioned into a more structured approach to concept study by aligning the activity with a graduate program in education. Our decision to do so was motivated by various drawbacks of the unstructured approach, some of which were mentioned in the previous chapter. In particular, we were concerned about

- *continuity* – The issue of regular participation proved problematic in the early group meetings. Even though these daylong meetings were scheduled during school hours, with substitute teachers hired to replace participants, it was not unusual for the group membership to vary by about 50% from one meeting to the next. The lack of continuity made it impossible to build up deeply shared understandings among many of the group members.

- *commitment* – Even among regular participants, the degree of commitment to concept studies varied considerably. Some participants sought out additional resources outside of meeting times in order to extend their explorations, studied in advanced, and adapted some of the concept study strategies for use in their classrooms. Others' efforts were contained in the scheduled meetings.

- *relevance* – Due to the inconsistencies in continuity and commitment among participating teachers, it was difficult to track the influence of the shared work on the teachers' practice. We found that we could not make consistent claims about the impact of the work on life in the participants' classrooms.

In other words, the decision to run a master's program at the University of

British Columbia, with a focus on concept study, was motivated largely by pragmatic concerns. Given our early concept study experiences and results, we proceeded on the hunch that longer-term engagement, with greater continuity and accountability, would have a substantial impact on teachers and their teaching. But we could not be sure of that. We made it clear to participants, prior to registration, that this was going to be a research-oriented program in mathematics education, for the purpose of investigating current research questions in M_4T.

The two-year program led to a Master's of Education degree. It consisted of 10 three-credit courses and a culminating project. As listed in Table 4.1, the coursework included five required "stock" courses and five electives. The two of us taught or co-taught most of the courses – and that meant that, although the five required courses were drawn from the regular offerings of the Department of Curriculum and Pedagogy, we infused them to the extent possible with topics and readings drawn from mathematics education. To do so, we used research articles in mathematics education as exemplars whenever we could. The final component of the program, the culminating project, was to be a ~50-page write-up of a concept study focused on a mathematical concept of personal and/or collective interest.

The program was designed for practicing teachers, and so the courses were offered outside of regular school hours – during the summer, on evenings and weekends, and/or online. Twenty-one teachers participated in the program. Among them were seven high school teachers, eight middle school teachers, and six primary school teachers. All of the teachers completed the program at the end of two years. Their unified progress through the program, made possible by the cohesiveness of the cohort structure, was a key ingredient in the creation and sustenance of a complex collective.

Collective Cognition

In the spirit of our introductory comments on complexity in Chapter 3, we organized the activities of the cohort around our understanding of the *collective* as a cognizing agent, as opposed to a *collection* of cognizing agents. This revised understanding afforded us, as researchers, both conceptual and pragmatic advantages. Most obviously, it allowed us to observe the "thinking" of the agent by noticing the interactions and prompts that triggered new possibilities and insights for the collective.

Observations and analyses of collective cognition have a fairly deep history in the field of mathematics education, even if the discussions have not been typically framed in terms of collective cognition of a social grouping. Such analyses have been provided by Bauerserfeld (1992), Cobb (1999),

The Math Teachers Know • Chapter 4

YEAR	TERM	REQUIREMENTS (DEPARTMENT COURSES)	OPTIONS (M4T-THEMED COURSES)
2009–2010	Summer		**Designing Mathematics Tasks** (taught by John Mason & Anne Watson) **Concept Analysis** (taught by John Mighton)
	Fall	**Research Methods in Education**	
	Winter	**Knowing, Learning, Teaching**	
	Spring	**Design-Based Research**	
2010–2011	Summer		**Lesson Study** **Concept Study** (taught with the assistance of Elizabeth Mowat)
	Fall	**Review of Curriculum Studies** (taught with the assistance of Steven Khan)	
	Winter		**Survey of Mathematics Education Research** (taught with the assistance of Steven Khan & Lissa D'Amour)
	Spring	**Writing Educational Research** (taught with the assistance of Shalini Khan)	

Table 4.1. The courses included in the M4T Master's Program at UBC

Kieran (2001), Kieren (2000), Sfard and Kieran (2001), and Zack and Graves (2001), to name a few. These researchers examined the development of interaction patterns and the emergences of social norms within mathematics classrooms.

This work has paralleled a broader discussion of "knowledge building" in the field of education (e.g., Scardamalia & Bereiter, 2003), a conceptual and pragmatic movement that is oriented toward the collaborative production of understandings that go beyond the level of the most knowledgeable individual in the learning group. Included among the core principles of knowledge building are the need to create an idea-rich environment, the convictions that all ideas are improvable, the requirement to gather and weigh evidence, and the centrality of participation. Although rooted in the social sciences literature, the knowledge-building frame is readily aligned with an emergentist frame – and, indeed, the word *emergent* appears frequently in the associated literature.

For example, and of particular relevance to our discussion of teachers' mathematical knowledge, Cobb (1999) has suggests a conceptual shift away from *mathematics as content* and toward *emergent terms*. As he explained, the "content metaphor entails the notion that mathematics is placed in the container of curriculum, which then serves as the primary vehicle for making it accessible to students" (p. 31). By contrast, when understood in emergent terms, mathematical ideas are "seen to emerge as the collective practices of the classroom community evolve" (p. 31). It is precisely this sense of mathematical emergence that we wished to enact and model with teachers in our concept study activities.

Use of the term *emergent* is common across many analyses and discussions, yet it is important to note that researchers frequently stop short of identifying classroom collectives or knowledge domains as learning systems. While they recognize that social norms, interaction patterns, and mathematical ideas emerge and evolve, there remains a strong tendency to describe these phenomena at the level of interacting agents, and not as properties of an emergent collective unity. Pragmatically speaking, we believe that explicit acknowledgment of collective cognition is vital if researchers wish to move from *describing* complex activity to *conditioning emergence* in the activities of complex unities. In other words, it is critical in efforts to give practical advice, both structural and procedural, to teachers.

In order to enact our understanding of the cohort as a complex collective as fully as possible, we made deliberate and explicit use of several principles of complex co-activity. These derived from earlier work on complexity in education (Davis & Simmt, 2003; Davis & Sumara, 2006), and included our paying close attention to the interplay between the ends of these structural pairings:

- internal diversity **v** internal redundancy – *Diversity* refers to structures that invite varied contributions, such as including teachers from all grade levels, and is seen as the root of the collective's intelligence. *Redundancy* refers to structures that ensure common understandings and communication, and is seen as the ground of the collective's coherence and robustness.

- limits **v** freedoms – *Limits* are well-defined boundaries that give structure and constrain activities from degenerating into aimlessness. *Freedoms* infuse activities with creativity and emergent possibilities. We explored the playful tension between the two by organizing activities less in terms of *must be done*, and more in terms of *what should be avoided*. Elsewhere these ideas have been captured in the notion of "enabling constraints" (Davis et al., 2008). These are structures for engagement that allow all creative options as long as they adhere to some clear guidelines.

As with the pairings introduced in Chapter 2, we link these ones with the "inclusive-or" of logic to remind that these dyads are complementarities rather than opposites. They are simultaneous, co-amplifying qualities of complex systems that can open up emergent possibilities.

Structures of Concept Study

The accounts of concept studies in the remainder of this chapter are all drawn from the group's engagements in the first required course, *Research Methods in Education*, which ran in the fall semester of 2009. The course was presented over six six-hour sessions, which ran roughly every other Saturday. The first half of each session was dedicated to the study of a specific research methodology, one of: phenomenology, hermeneutics, ethnography, statistics-based inquiry, design-based research, or action research. The afternoons were spent in discussions of how the methodology might be used to investigate the issue of mathematics teachers' disciplinary knowledge. The persistent focus on the M_4T question allowed us to acquaint the cohort with some of the pertinent research literature.

Another intended outcome of the afternoon sessions was to assist the participants in starting their own concept study work by familiarizing them with the structures of concept study that had emerged in the work of other groups. We chose to focus on four "emphases" that had proven productive, in different groups and across different topics, for the collective elaboration of mathematical concepts (Davis & Renert, 2009). We titled these emphases *realizations, landscapes, entailments,* and *blends*.

Before describing the emphases in greater detail, however, it is important to note that we do not offer them as "steps" or "levels" in a linear process.

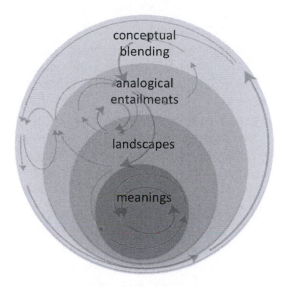

Figure 4.2. A visual metaphor to depict the relationships among concept study emphases

Rather, we see them as interpretive strategies that are always simultaneously present, alongside others that are yet to be noticed or made explicit. Moreover, these emphases and the strategies they entail are not to be implemented rigidly. Rather, we see their interpretive potentials as nested and inseparable – more akin to the visual representation in Figure 4.2 than to a ladder or a staircase.

Another important caveat before we delve into the emphases is that it is impossible to report on concept study work without presenting some of the "findings" of the group in detail. In presenting these details, we in no way intend to suggest that *these* teachers identified what *all* teachers should know. In fact, we would be horrified if the results of this or any other concept study were to find their way into a textbook, or onto a syllabus for an M_4T course. Rather, the vital component of our reporting resides in its descriptions of the structures and dynamics of the engagements. We believe that *all* mathematics teachers would benefit from engaging in concept study activities with similar structures and dynamics.

As we argued previously, while an important body of knowledge exists that teachers can and should master, an open disposition towards emergent mathematics in pedagogical settings remains a critical element of M_4T. In our work, we have repeatedly witnessed the interpretive emphases that we are about to describe promoting such openness among teachers, both to their own and to others' mathematical understandings.

Emphasis 1: Realizations

We borrow the term *realizations* from Sfard (2008) and use it refer to what we have heretofore been describing as "meanings," "interpretations," and "instantiations." Briefly, as Sfard explains, the notion of realizations is used to collect all manner of associations that a learner might draw on and connect in efforts to make sense of a mathematical construct. More precisely, a realization of a signifier *S* refers to "a perceptually accessible object that may be operated upon in the attempt to produce or substantiate narratives about *S*" (p. 154). The distinction between a signifier and a realization is often blurred, as mathematical realizations can often be used as signifiers and realized further. Among many possible elements, realizations might draw on:

- formal definitions (e.g., multiplication is repeated grouping)
- algorithms (e.g., perform multiplication by adding repeatedly)
- metaphors (e.g., multiplication as scaling)
- images (e.g., multiplication illustrated as hopping along a number line)
- applications (e.g., multiplication used to calculate area)
- gestures (e.g., multiplication gestured in a step-wise upward motion).

To be clear, the assertion and assumption here is not that any particular realization is right, wrong, adequate, or insufficient. It is that personal understanding of a mathematical concept is an emergent form, arising in the complex weaves of such experiential and conceptual elements. With regard to teachers' knowledge, we offer that realizations are the "objects" or "agents" of the complex system of mathematics for teaching.

It's important to note that the sorts of realizations listed in the previous paragraph vary considerably in nature. A more nuanced discussion might group these entries into subcategories. To this end, Bruner (1966) offered a popularly cited rubric, in distinguishing *enactive* (action-based), *iconic* (image-based) and *symbolic* (language-based) representations, an illustration of which we present in Table 4.2. An essential element of Bruner's typology is the assumption that a learner goes through a sequence of transitions to move from an operational (i.e., action-based) concept to a symbolic (language-based) concept, and this point is reflected in the common belief among mathematics educators that operational conceptions are, for most, the first stage in the learning of a mathematical concept (see, e.g., Sfard, 1991). An important, related claim is that "processes performed on certain abstract objects turn into new objects that serve as inputs to higher level processes" (Even & Tirosh, 2002, p. 219). Building on this insight, others have offered much more fine-grained models of the growth of mathematical understanding, in some cases involving many more categories of representation and engagement (see, e.g., Ainsworth, 1999; Ainsworth & van Labeke, 2004; Lesh, Post, & Behr, 1987; Pirie & Kieren, 1994; Presmeg, 1986).

Representation Type	The product of 2 and 3 might be experienced/represented as ...
enactive (action-based)	... actually hopping a distance of two tiles three times, or dropping three pairs of marbles in a marked area
iconic (image-based)	
symbolic (language-based)	"two multiplied by three" is "two times three" or "2 + 2 + 2"

Table 4.2. Bruner's Typology of Representations, illustrated with the example of "2 × 3"

We mention this work because, in offering our undifferentiated list of realizations a few paragraphs ago, we do not want to suggest that we are ignoring or conflating important variations among representations encountered by learners. Rather, our main reason in leaving them "amassed" is to signal how they are typically experienced by learners and invoked by teachers: all at once, without categorization.

It's also important to highlight that this emphasis on realizations is supported by a deep body of research. The role of varied representations in conceptual understanding has long been a prominent topic of discussion and research in the mathematics education community (e.g., Fischbein, 1989, 1993; Goldin & Janvier, 1998; Janvier, 1987), and rightly so. Varied realizations appear to be very important in early mathematics learning – there is at least circumstantial evidence that it makes a difference to provide learners with diverse interpretations. For example, teachers in high-performing jurisdictions, such as Hong Kong and Japan, were roughly twice as likely as U.S. teachers to invoke varied interpretations of concepts (Hiebert et al., 2003). Analogies can be useful, provided that novices have access to sustained interpretive assistance (Richland et al., 2006; Zook, 1991). But despite considerable research on realizations of grade-school mathematics concepts (e.g., English, 1997; Lakoff & Núñez, 2000; Presmeg, 1986), the topic has not been systematically incorporated into teacher preparation.

The process of collectively identifying realizations, and their related signifiers, is neither linear nor obvious. Each knower holds and utilizes a personal set of realizations. Some of these are common to all participants, while others are idiosyncratic or shared by only a few. Moreover, realizations are not fixed. They evolve through the process of learning. Not only they become more numerous, some earlier ones are discarded or expanded when new contexts arise. Some realizations (e.g., "multiplication is repeated addition") may be so well rehearsed that they may overtake or rule out other interpretive possibilities.

To circumvent the tendency to go directly to well-rehearsed definitions, we invited the teachers to explore how the concept of multiplication is introduced, taken up, applied, and/or elaborated at different grade levels. We also asked the participants to identify problems that learners encounter as they study multiplication. These opening prompts contributed to a rich set of realizations. To encourage even deeper exploration, we cautioned that commonplace definitions of multiplication, such as repeated addition, might actually be barriers to understanding multiplication when dealing with fractions, signed integers, and other number systems.

In order to promote cross-grade diversity, we asked the participants to pay particular attention to colleagues who taught at different grade levels, with the added proviso of "the greater the difference, the better." Finally, we advised from the start that there was "no right answer". We had to repeat this point several times during this meeting as a number of participants wondered aloud if they were getting close to "what was expected," or voiced frustration that the task was "too difficult."

The collected list of realizations, assembled after small groups worked for 30 minutes, is presented in the Figure 4.3.

It bears noting that there are some striking similarities and and some significant differences between this list and those generated by other groups (e.g., Davis & Renert, 2009; Davis & Simmt, 2006). We have observed remarkable stability across groups of teachers – pre-service and practicing, novice and experienced, grade-specific and cross-grade – in the range and quantity of interpretations generated. In every context so far, participants have generated about a dozen realizations of multiplication, of which "grouping" and "repeated addition" are always the first. (Of course, such stability is not surprising. To date, participants have been drawn from the same edu-ecosystem.)

Also notable with this group was their decision to include a "lingering worries" section. It signaled to us a high level of awareness of the purpose of the task. The three worries listed by the teachers indicated that they were aware of the need to help students make sense of multiplication, while at the same time they acknowledged that they did not have ready interpretations to deal with these specific instances. The teachers would later return to these worries in subsequent sessions, and would discover that they had more to do with *not having immediate conscious access to what is known* than with *not knowing*.

Some Realizations of Multiplication

- grouping process
- repeated addition
- times-ing
- expanding (i.e., distributing across factors; e.g., how can you write out $(x + 3)$, $(x + 2)$ times?)
- scaling
- repeated measures
- making areas (continuous)
- making arrays (discrete)
- proportional/steady increase/slope/rise
- splitting, folding, branching, sharing, and other 'dividings' (i.e., multipyings)
- skip counting (jumping along a number line)
- transformations
- stretching/compressing a number line

Lingering Worries:

- none of these on their own seems to do much to illuminate -2×-3
- what's going on when units are introduced – e.g., 3 g/L × 5 L
- most of the entries rely on or suggest a linear or rectilinear basis/image of multiplication, which might work through middle school, but won't stretch across some topics encountered in and beyond high school

Figure 4.3. Teacher-generated realizations of multiplication, and related concerns

To re-emphasize we do not regard individual realizations as discrete or isolated components within a concept. Rather, we see realizations as interacting elements within the evolving system of teachers' knowledge of mathematics. Since realizations evolve through the process of learning, it might be productive to construe aspects of mathematics learning as the evolution of networks of realizations.

Emphasis 2: Landscapes

There are dramatic differences of conceptual worth among realizations. Some can reach across most contexts in which a learner might encounter a concept – for example, and array/area-making interpretation of multiplication can be useful for multiplying whole numbers, fractions, and algebraic expressions. Others are situation-specific or perhaps learner-specific – such as repeated addition, which has limited interpretive value beyond applications involving whole numbers. For us, this insight invited the question of how realizations relate to one another, which in turn compelled a strategy to organize and contrast the entries on assembled lists of realizations. Because past efforts to organize realizations often resulted in two-dimensional tables and maps, we selected the word *landscapes* to refer to this emphasis of concept study. Briefly, a landscape is a macro-level map, whereas a realization is a micro-level snapshot, of a concept.

As illustrated through this chapter's opening anecdote, the landscape emphasis was invented by a group of teachers who were frustrated with the incoherence of a raw list of interpretations for a concept. Their efforts to create landscapes were inventive and far-ranging. As noted in the anecdote, they made use of a wide array of organizers, including grade level, types of applications, underlying metaphor, continuous/discrete, conceptual/procedural, and enactive/iconic/symbolic. In particular, grade level proved to be a very useful organizer.

Within the research methods course, the landscapes activity took place a few weeks after the initial realizations exercise. Recognizing the importance of grade-level orderings from past work, this time we invited the participants to identify "break points" – that is, instances in the K–12-mathematics curriculum that compel significant shifts in understandings of multiplication. One result was a grade-indexed listing of realizations and curriculum topics (Figure 4.4).

It was at this point that the value of the group's shared readings became apparent. One of the required readings in the course was Greer's (1994) paper on the meanings of multiplication and division. In it, he noted that

Grade	Context(s)/Interpretation(s)
K	grouping
1	skip-counting
2	
3	repeated addition (whole numbers) multiple folds (fractions)
4	double-digit multiplication; perimeter; inverse of division
5	operations on decimals
6	ratios and proportions
7	
8	operations on fractions; operations on integers
9	exponentiation; polynomials; rational expressions
10	radicals; irrational; negative exponents; slope; direct and indirect variation
11	complex numbers; functions; transformations (expansions; compressions)
12	permutations and combinations; logarithms

Figure 4.4. A preliminary grade-indexed listing of topics in multiplication

many different criteria could be used to organize this sort of information. Their familiarity with the text made it natural for the group to move beyond grade-level sorting to seek other distinctions that might be used as axes in a mapping of realizations. Many distinctions were proposed and, as with the realizations emphasis, the suggestions were both similar to those generated by other groups and unique to this setting. Criteria included: grounding metaphors, underlying image (e.g., number line; grid; area, graph), dimension(s) of underlying image (0, 1, 2, 2+), types of factors (i.e., discrete/continuous), curriculum topics, processes versus objects, keywords (e.g., "by," "of," "times"), applications, factors with/without units, and errors/misconceptions.

The participants then proceeded to construct maps and tables on the basis of the different criteria. The products of their work were posted and compared. It was a veritable gallery of diagrams and maps. What struck us most is how different they looked from one another. We then suggested that the group proceed to create a collective mapping on the basis of the "most useful" categories and distinctions. After some discussion and voting, the teachers selected the categories of grade-level and grounding metaphor. The mapping presented in Figure 4.5, in which Lakoff and Núñez's (2000) notion of "grounding metaphor"[1] is used to organize constructs along the horizontal axis, is one of more than a dozen distinct landscapes created by the group using different axes.

We view this exercise not only as descriptive but also as creative. For us it was a clear instance of the emergence of novel mathematical knowledge. The collective generated insights that were not available to any its members prior to the engagement, but that depended entirely on the combined knowledge of all participants. That is, the knowledge *was there*, but it was inaccessible in its entirety to any individual, partly because of its largely tacit nature and partly because aspects of it were held by different members of the group. These partly tacit, partly distributed insights were given voice and collected in the discussion that followed. They included:

- Multiplication is introduced informally (but not explicitly) in terms of motion as early-grade teachers skip count (in particular, by 2, 3, 5, and 10). Frequently, skip counting is accompanied by tracing out the movement on number lines and in number charts. An implicit emphasis

---

1. Lakoff and Núñez's four grounding metaphors of arithmetic, as listed across the top of Figure 4.5, are ARITHMETIC AS OBJECT COLLECTION, ARITHMETIC AS OBJECT CONSTRUCTION, the MEASURING STICK METAPHOR, and ARITHMETIC AS MOTION ALONG A PATH. Their descriptions of these metaphors are detailed and well-illustrated. For our immediate purposes, it suffices to say that these metaphors align with four distinct interpretations of number – respectively, NUMBER AS COUNT (cardinality), NUMBER AS MAGNITUDE, NUMBER AS MEASURE, and NUMBER AS POSITION.

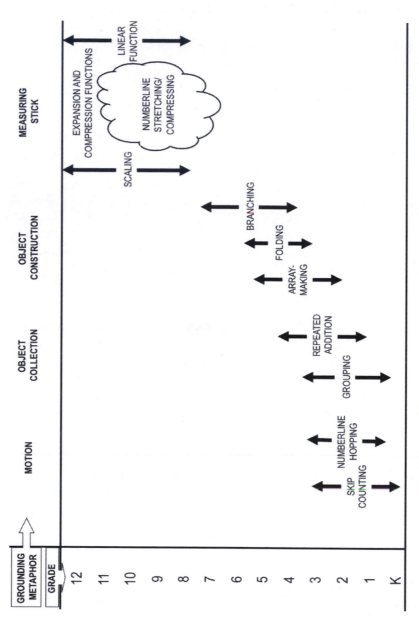

Figure 4.5. A collective mapping of some of the realizations

on MOTION precedes an explicit emphasis on the constructive activities of grouping and repeated addition.

- The movement across categories of interpretation in different grade levels seems to be fluid and deeply connected; yet it is rarely made explicit. For example, the sequence of "SKIP-COUNTING ⇨ NUMBER-LINE HOPPING ⇨ GROUPING ⇨ REPEATED ADDITION" had not been previously noticed by any of the participants. When this sequence was rendered explicit by the investigation, the participants agreed that it was an "easy" and "natural" progression. The qualities of easy and natural were in turn critiqued as rooted in well-rehearsed sequences of experience and instruction (i.e., natural to participants' histories in the Canadian edu-ecosystem, and easy relative to practiced applications).

- For the purposes of school mathematics, line-based (and, in particular, number-line based) interpretations of multiplication appear to be the most flexible and powerful. In particular, with appropriate accommodations, multiplication as STRETCHING/COMPRESSING A NUMBER LINE[2] can be made to reach across all number systems encountered in grade school. (Note that this interpretation has been argued to be the most critical for more advanced mathematics as well; see Mazur, 2003.)

- There are also pedagogical moments in which elaborations are important yet not intuitively obvious. For example, there appears to be a major conceptual leap when one shifts from OBJECT COLLECTION interpretations (which are well suited for whole-number applications) to OBJECT CONSTRUCTION interpretations (which are useful for continuous applications, including other number systems and algebra).

The insight that perhaps held the most impact for the participants had to do with the large number of realizations of multiplication introduced in the early grades (K–6). In our follow-up discussions, group members wondered if this "explosion" of meanings might contribute to the common confession, "I was good at math until grade 6" or, even more troubling, "I liked math until grade 6." A young learner trying to understand a central mathematical concept, such as multiplication, and faced with increasing complication

---

2. Briefly, and as touched on in the next section, the MULTIPLICATION AS STRETCHING/COMPRESSING A NUMBER LINE realization involves interpreting one factor as a scaling ratio, the second factor as an originating position, and the product as a final position, as illustrated in the diagram to the right.

Among the advantages of this realization is the ready application to integers and rational numbers. See Mazur (2003) for a more detailed discussion.

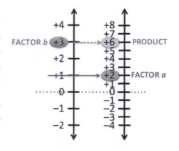

and little interpretive assistance, might begin to dislike the subject matter. The alternatives to deep mathematical understanding – which include rote memorization and routinized application – do not engender interest in and love of the subject matter.

Emphasis 3: Entailments

As discussed in Chapter 3, each realization of a concept carries a set of logical implications and entailments. For example, as illustrated in Figure 4.6, as one moves among realizations of multiplication, meanings of multiplier, multiplicand, and product also shift. Notably, the implicit actions underlying these explicit images vary dramatically and include clustering items together, extending in two dimensions, shrinking, and rotating.

The intention of this third emphasis is thus to study how different realizations shape the understanding of related mathematical concepts (e.g., how the realization of multiplication as NUMBER-LINE HOPPING give form to an understanding of the commutative property of multiplication). In the process of exploring different entailments, participants are compelled to consider mathematical concepts afresh and not only in well-rehearsed ways. Some surprises emerge.

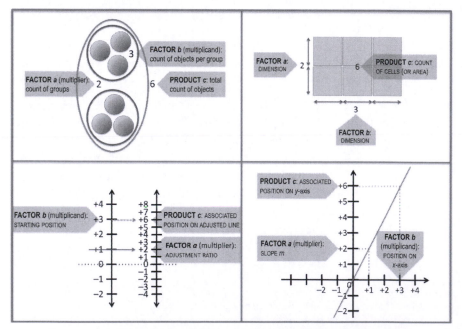

Figure 4.6. Entailments for multipliers, multiplicands, and products of four realizations of multiplication (respectively: grouping, array/area-making, number-line expansion/compression, and slope)

If multiplying is ...	... a factor is ...	... a product is ...
REPEATED ADDITION		
REPEATED GROUPING		

Figure 4.7. A "starter" entailments chart

As mentioned, the "entailments chart" tool used in this emphasis originated in the "Is 1 prime?" study (see Chapter 3). We introduced it within the research methods course, by showing the group the small matrix in Figure 4.7.

Before looking at some of the detail of the group's work, we would like to point out that the modes of thinking required to fill in an entailments chart highlight an important distinction between *unpacking* and *substructing*, which were mentioned in the previous chapter. We believe that the mode of engagement used in identifying realizations is appropriately called "unpacking." It involves looking for associations that are already there and are familiar – so familiar that they can be difficult to notice. In contrast, the entailments of different conceptual associations can prove to be surprisingly unfamiliar. Working out the entailments can be tedious and frustrating. This deliberate substructing work pushes in the opposite direction from which the automatic side of our brains (i.e., System 1) wants to go. Recall that System 1 wants to streamline, to be efficient, and it accomplishes this by forgetting how it achieved its expertise. Put differently, the entailments chart is inviting teachers to be novices with regard to their assumed-to-be expert knowledge. We believe that it's this new positioning of teachers as novices that gives rise to the difficulties and resistances encountered during this emphasis. Figure 4.8 presents a consolidation of work that was conducted in small groups. Notable in this table is the "commutativity" column, which was not specified by us but arose out of the conversations. Before examining the chart, we would invite you to pause for a moment to think through some entailments on your own. For example, what is the meaning of commutativity when multiplication is understood as a skip-counting process? Or as scaling? These sorts of questions may not be as straightforward to answer as they seem at first.

This is exactly what the group members reported. Many of their emergent insights were accompanied by expressions of surprise and confusion.

If multiplying is ...	... a factor is ...	... a product is ...	... commutativity is ...	... a prime is ... (necessary conditions, but not sufficient)
REPEATED ADDITION	ADDEND or NUMBER OF ADDENDS (2 × 3: 2 added to itself 3 times or vice versa)	a SUM	$2 + 2 + 2 = 3 + 3$	sum of ones
REPEATED GROUPING	NUMBER OF GROUPS or NUMBER OF ELEMENTS IN EACH GROUP	a SUM: total number of all the elements in the groups (cardinality of the set)	2 groups of 3 = 3 groups of 2	one group, or one element in each group
MAKING A GRID OR RECTANGULAR ARRAY	DIMENSION: number of rows (number in each column) and number of columns (number in each row)	NUMBER OF CELLS	90-DEGREE ROTATION (a 2-by-3 grid has the same number of cells as a 3-by-2)	one of the dimensions has to be 1
SKIP COUNTING	SIZE OF THE JUMP and NUMBER OF JUMPS	END DESTINATION (the last number you land on)	a jumps of distance b lands you in the same place as b jumps of length a	must make only a single jump or jump one space at a time
SCALING	SCALE FACTOR and ORIGINAL MEASURE	MEASURE OF THE FINAL MAGNIFICATION/REDUCTION	size a scaled by a factor of b gives the same result as size b scaled by a factor of size a	when a magnification/reduction can only be reached in unit increments or directly

AREA GENERATION	DIMENSIONS (lengths and widths)	AREA	90°-ROTATION: $lw = wl$	one dimension must be 1
NUMBER-LINE STRETCHING AND COMPRESSING	SCALE FACTOR and STARTING POSITION ON UNALTERED NUMBER LINE	CORRESPONDING POSITION ON STRETCHED/ COMPRESSED NUMBER LINE	If c corresponds to point a when line is scaled by b, it will correspond to point b when scalar is a	to get to point c, you must EITHER start at 1 with a scalar of c OR vice versa
FOLDING	NUMBER OF HORIZONTAL AND VERTICAL DIVISIONS (made by the folds)	NUMBER OF LAYERS	folding into a layers, then into b layers gives the same number of layers as b first, then a	can only be folded directly using $a - 1$ folds
BRANCHING	NUMBER OF STEMS and NUMBER OF BRANCHES PER STEM	TOTAL NUMBER OF BRANCHES AT THE LAST LEVEL	a branches of b stems has the same product as b branches of a stems	must either have 1 stem or 1 branch/stem
LINEAR FUNCTION $y = mx$	SLOPE and x-COORDINATE	y-COORDINATE	If $y = c$ when $m = a$ and $x = b$, then $y = c$ when $m = b$ and $x = a$.	To get to the desired y-coordinate c, either $m = c$ and $x = 1$ or $m = 1$ and $x = c$.

Figure 4.8. Some analogical implications of different realizations of multiplication

(Note: UPPERCASE is used to flag realizations – of both multiplication and associated concepts)

For instance, the participants noted how obvious the commutative property appears when an array/area interpretations of multiplication is used and how obscure it is when a scaling interpretation is used. One participant commented, "I get the idea of the communicative property – flipping the factors doesn't change the product – but I'm realizing I don't get what's under it."

As anticipated based on work with earlier concept-study groups, the notion of *"prime"* provided similar surprises. In particular, participants lingered over their noticing that realizations rooted in continuous phenomena (e.g., number lines, areas, and linear functions) were not well suited to understanding of primeness. This insight led to a discussion of how teachers "choose" interpretations without thinking – that is, they select realizations that fit well with specific aspects of a concept, without consciously reflecting on the entailments of other realizations that their students might be using.

The participants then moved on to discuss what implicit criteria teachers might be using to select realizations. Possibilities included:

- mathematical appropriateness;
- sufficiency for the situation at hand; and
- familiarity with the realizations.

After some further discussion, participants agreed that the practice of teaching must entail careful and systematic attention to the choice of realizations. To this end, other vital criteria for selection were suggested:

- conceptual reach;
- potential for elaboration; and
- transparency.

The session closed with the reflection that teachers' capacity to move fluidly among realizations was not so much a matter of "having a firm handle" on those realizations, but rather a consequence of losing the ability to differentiate among them.

Emphasis 4: Blends

The three emphases described so far – realizations, landscapes, and entailments – are focused mainly on making fine-grained distinctions among interpretations. Not surprisingly, while the participating teachers showed strong interest in these emphases, they also voiced some frustrations as the shared work unfolded. Multiplication is, after all, a mathematically coherent concept, not an assemblage of images and implications that can be laid out in discrete pieces. The blending emphasis is intended to address this concern. It is about seeking out meta-level coherences by exploring the deep connections among realizations and by assembling those realizations into grander, more encompassing interpretations that yield further emergent interpretive possibilities.

Tying back to the earlier mention of Bruner's typology of realizations (i.e., action-based *enactive*, image-based *iconic*, and language-based *symbolic*), the move here might be described in terms of a more deliberate engagement with the formal processes of mathematical thinking, following Tall (2004). He identified three distinct ways of thinking about mathematics:

- conceptual (embodiment);
- proceptual (symbolism); and
- axiomatic (formalism).

All three worlds make use of varied sorts of enactive, iconic, and symbolic realizations – but the manner if which such realizations are engaged differs from one to the next. In Tall's terms, the above three categories constitute more than a sequence of increasingly sophisticated modes of operation. Rather, they describe different "worlds of mathematics," each with a distinct frame of reference, its own set of conventions, and its own grander systemic character.

So far in this chapter, the three emphases we have addressed (i.e., realizations, landscapes, and entailments) operate mainly in the first two of Tall's worlds of mathematics. We see the emphasis on conceptual blends as a more deliberate move into a formal, axiomatic space. It is oriented by a recognition of the necessary movement toward formulating mathematics in terms of set-theoretic axioms. In essence, it represents a shift in emphasis from multiple (and potentially disjointed) meanings toward coherent and encompassing definitions.

For this emphasis, we drew principally on cognitive science research into conceptual blends (Fauconnier & Turner, 2003; Seufert, 2003), which examined the emergence of new and more powerful discursive objects through combinations and mash-ups of existing ones. In particular, following diSessa (2004), we introduced conceptual blends to the cohort in terms of *metarepresentations*. As diSessa described, metarepresentational skills comprise "modifying and combining representations, and ... selecting appropriate representations" (p. 296), subcomponents of which include inventing and designing new representations, comparing and critiquing them, applying and explaining them, and learning new representations quickly.

In previous concept studies, a few blends/metarepresentations arose spontaneously (Davis, 2008; Davis & Renert, 2009; Davis & Simmt, 2006). However, in this group, we tried to condition the emergence of blends in a more deliberate manner. We found out that *causing* emergent insight was impossible. This is not to say that the time set aside for this emphasis was unproductive. We had decided in advance that, should explorations stall, we would introduce metarepresentations from other concept studies. The cohort's engagements around them produced some interesting emergent insights.

The first blend to be introduced was from a concept study reported by Davis & Simmt (2006), in which the combination of a area-based image and a grid-based algorithm was used to highlight some connections among standard algorithms for multiplying multi-digit whole, decimal fractions, mixed numbers, and binomials (Figure 4.9).

The second was from a study reported by Davis & Renert (2009), through which participants developed an interpretation that brought together multiplication as NUMBER-LINE STRETCHING/COMPRESSING, as SCALING, as a LINEAR FUNCTION (Figure 4.10). These realizations can be blended together by noticing that all involve a pair of number lines. In the NUMBER-LINE STRETCHING/COMPRESSING realization, the two lines are parallel (and can be linked by tracing across perpendicular lines); in the LINEAR FUNCTION interpretation, the two lines are perpendicular to one another (and can be linked by tracing up to and across from sloped lines). As described elsewhere, the blended product that emerged had several productive features. Among them was a new way to understand integer multiplication in terms of a number line that inverts as it compresses through zero.[3]

The surrounding conversation focused mainly on the manner in which some of the previously identified realizations did or did not fit within these metarepresentations. Another prominent theme was the extent to which the two metarepresentations might be reconciled into a unifying blend.[4]

---

3. The explanation that the group developed is quite lengthy, and so we haven't included it in its entirety here. Another reason for skipping most of it is that, in our experience, it's usually better to leave the task of making sense of things to participants. Two hints have proven to be particularly useful, however:

- the similarity between initial realizations (i.e., NUMBER-LINE STRETCHING/COMPRESSING SCALING and LINEAR FUNCTION) might be highlighted by mentioning that the starting number line of the former (i.e., the "×1" line) maps onto the x-axis of the latter, and

- the target number lines of the former (i.e., the "×½" and "×2" lines) can be imaginatively "made to fit" onto the y-axis of the latter when the grid is "rotated in a third dimention" either about the x-axis (for multipliers greater than 1 or less than −1) or the y-axis (for multipliers between −1 and 1) in a manner that "locates" the function $y = mx$ or the rotated graph atop the $y = 1x$ of the original graph.

4. Once again, we don't feel a complete explanation is appropriate here (for the same reasons given in the previous footnote). However, a useful hint for the concept-study group to help reconcile the metarepresentations presented in Figures 4.9 and 4.10 was the pattern of products on the "multiplication grid" to the right (in which the smaller numbers represent the products of the coordinates for each point).

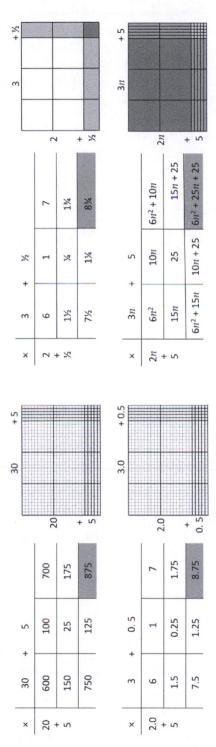

Figure 4.9. A grid-based metarepresentation that highlights the similarities of multiplicative processes involving additive multiplicands

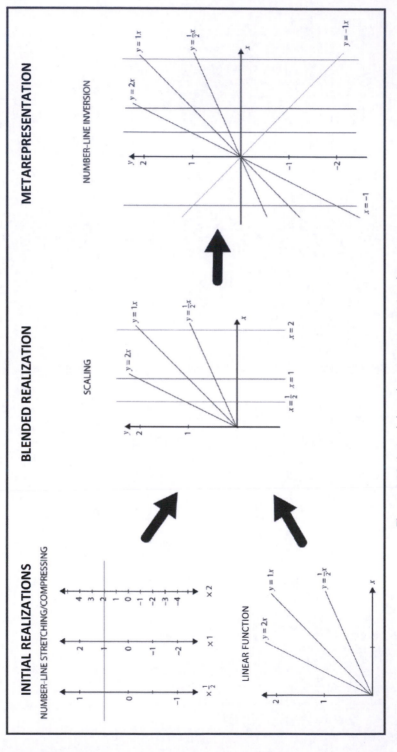

Figure 4.10. A graph-based metarepresentation that blends linear models of multiplication

More generally, discussions revolved around the conceptual novelty that can emerge in metaprepresentations, and how, in turn, such metarepresentations might eventually be blended into even more encompassing, emergent interpretations. Although our hope that novel metarepresentations of multiplication would emerge in this cohort was not met, the participants did benefit by expanding their awareness of the possibilities inherent in mathematical webs of associations. Within the context of the master's program, this emphasis achieved the goal that was intended: to highlight the importance (and complexity) of meta-awarenesses of mathematical concepts, so that the participating teachers might include them in their own inquiries into topics of their own choosing.

Where Are we Going with This?

> The *determinant* of a matrix is an oriented volume of the parallelepiped whose edges are its columns. If the students are told this secret (which is carefully hidden in the purified algebraic education), then the whole theory of determinants becomes a clear chapter of the theory of poly-linear forms. If determinants are defined otherwise, then any sensible person will forever hate all the determinants, Jacobians and the implicit function theorem. (Arnold, 1997)

Admittedly, a discussion of the determinant of a matrix as ORIENTED VOLUME OF THE PARALLELEPIPED is a long way removed from this chapter's discussion of multiplication. In terms of M_4T, however, there is some important common ground. The simple (yet utterly complex) point is that mathematical concepts are comprehensible, but that their comprehensibility demands continuous elaborative work – and this is true for both the student and the teacher, and for both the most elementary and the most advanced of mathematical concepts.

Attention to the ongoing elaboration of mathematics is important for at least two reasons. First, our concept studies have shown time and again that virtually every concept introduced in the early grades – number, order, equivalence, addition, etc. – undergoes regular conceptual elaborations and shifts as students move through new topics and are introduced to new applications. Second, as the example of multiplication illustrates, these elaborations are never simply accumulative. Rather, they are always transformative, and compel some sort of revision to the network of associations that constitutes one's "understanding" of a concept.

On that count, we re-emphasize our conviction that teachers' disciplinary knowledge of mathematics cannot be reduced to a body of knowledge that might be catalogued, instructed, and tested. While it may include some such components, the more critical element of M_4T is the open disposition toward the evolution of concepts. Teachers must have more than an access to an established domain of knowledge; they must have means to unpack, interrogate, and elaborate – that is, to substruct – their mathematics.

As illustrated in our comparison of the concept studies of multiplication by two different groups, substructing activities need not home in on the same sets of understandings. There is room for diversity of interpretation and creativity. To that end, the emphases/strategies introduced in this chapter – realizations, landscapes, entailments, and blends – are intended only as "potentially useful devices." This point was brought home to us regularly as we worked through later concept studies (as we recount in Chapters 5 and 6). Working on their own chosen concept, the teacher-participants found some utility in our pre-structured emphases. However, for the most part, their joint inquiries unfolded in more organic ways, especially as the explorations and conversations pulled toward the everyday, pragmatic realities of teaching.

PEDAGOGICAL PROBLEM SOLVING

THE EMERGENCE OF A COMMUNITY OF EXPERTS

— In brief …

This chapter recounts the evolution of the concept study methodology into more pedagogically efficacious forms that occurred when teacher-participants selected topics of personal interest and undertook to conduct their own concept studies. A new emphasis emerged when the teachers then met to investigate problems encountered in their teaching. Participants were able to combine the expertise gleaned from their diverse concept studies to explore solutions to these problems. This type of concept study engagement, that we call "pedagogical problem solving" is, in general, more complex than the study of solitary concepts.

Pedagogical Problem Solving

[VOICES: BRENT AND MOSHE]

In the winter term of 2010, on the heels of their course on research methods in education, participants in the UBC master's cohort began to work on their own concept studies. After organizing into self-selected groups of one to five, they chose topics of study that included subtraction, circles, zero, factoring, fractions, and division.

For some of these topics, the emphases introduced in the research methods course — realizations, landscapes, entailments, and blends — worked well and the investigations progressed smoothly through the term. For others, these emphases were less useful and group members soon found themselves inventing other strategies for substructing concepts. As an example, more than one group thought it would be useful to make sense of common errors and misconceptions of students around the concepts under investigation.

Noticing the very practical orientation of this emerging focus, we decided to challenge the participants to bring their diverse and newly developed expertise to bear on a question derived from pedagogical practice. We wondered if doing so might occasion a productive exploratory conversation that would produce useful insights for teaching. At the start of the next course, which began during the summer break, we asked the teachers to suggest questions, related to the concepts they had been investigating, that have proven challenging for their students in the past. Among the questions brought forward were:

- *Is ∞ a number?*
- *What does it mean to divide by zero?*
- *What is the difference between* undefined *and* indefinite*?*

In an attempt to pull together some of these questions, we began by asking, "What's 5/0?" We invited the participants to analyze this question by referring to the insights gained hitherto from their concept studies.

Right from the beginning of the activity, it became clear that the participants would have to contend with the terms undefined, indeterminate, *and* infinite, *as these are answers that are regularly advanced to the question "What's 5/0?" So before proceeding, we offer three formal definitions, taken from Wolfram MathWorld:*

Indeterminate: *A mathematical expression can … be said to be indeterminate if it is not definitively or precisely determined.*

Infinite: *Greater than any assignable quantity of the sort in question.*

Undefined: *An expression in mathematics which does not have meaning and so which is not assigned an interpretation.*[1]

From a linguistic perspective, these three terms have similar original meanings, even though they have very different definitions.[2] *All three carry senses of openness, unending, and unspecifiability, which might well contribute to conflations and confusions among learners. As we found out during the investigation, depending on the interpretations that a learner attaches to the concepts involved, 5/0 can be legitimately construed as indeterminate, infinite, or undefined. We offer three interpretations from among the many that arose during the concept study.*

There are many ways to interpret "5" (e.g., as a discrete count, a position on a continuous number line, a comparison value, etc.), the slash "/" (e.g., as divided by, inverse of, out of, per, in, to, etc.), and "0" (e.g., as nothing, starting position, center point, stationary, etc.). We invited the participants to bring up compelling interpretations that draw on findings from their own concept studies.

Different groups contributed different interpretations. The division group suggested that the class interprets division as sharing, inverse of multiplication, and

---

1. Retrieved 2012 July 18.
2. According to the *Oxford English Dictionary*, *indeterminate* is drawn from the Latin *in-* ("not") + *de-* ("off") + *terminare* ("to mark the end or boundary"); *infinite* derives from the Latin *in-* ("not") + *finis* ("end"); *undefined* combines *un-* ("not") + *definire* ("to limit, determine" – which, in turn derives from the root *finis*).

repeated subtraction. If the division is interpreted as sharing, and the 5 and 0 are interpreted as quantities, then 5/0 can be understood either quotitively (i.e., how many groups of 0 are in 5?) or partitively (i.e., what is the size of each group when 5 is partitioned into 0 groups?). The class decided that the former "makes sense" while the latter appears to contain a contradiction (i.e., if there are no groups, it makes no sense to ask about the size of each group). Accordingly, one can argue that 5/0 is

> infinite, *with reference to the quotitive interpretation, because there are clearly a boundless number of size-0 groups can be drawn from a group of 5 — i.e., the quotient is "greater than any assignable quantity of the sort in question."*

> undefined, *with reference to the partitive interpretation, because the expression is meaningless and cannot be assigned a sensible interpretation.*

Similar responses were obtained when division was interpreted as repeated subtraction.

The situation did not get any clearer when the factoring group weighed in with its interpretation. As that group saw things, the quotient 5/0 is equivalent to the missing factor, n, in 0 × n = 5. There is no such factor, and so 5/0 must be

> indeterminate — *because, although the expression can be assigned a sensible interpretation, it cannot be definitively or precisely determined.*

Another group approached the question by focusing on the meanings of zero. It also found distinct rationales for all three responses. As the possible perspectives kept mounting, it became clear that the teachers' newly forged "community of experts" has given rise to a vast array of interpretive possibilities.

"Pedagogical problem solving," an emergent fifth emphasis in concept study, represents a move into some of the more complex processes entailed in teaching mathematics, one that goes beyond the study of discrete concepts. This emphasis is, in a very deep sense, a site of the *real mathematical work of teachers*, as it focuses on the mathematical problems that they encounter daily and that are specific to their profession. Unlike earlier emphases, which tended to artificially circumscribe mathematical concepts and meanings in ways that are not likely to be implemented in mathematics classrooms, this new emphasis is developed around the actual questions that meaning-seeking learners ask.

Pedagogical problem solving capitalizes on the interpretive potentials that arise collectively when teachers draw on various instances of individual expertise in order to broach perplexing problems of shared interest. It situates the enterprise of concept study in the everyday complexities of mathematics teaching and problem solving, where multiple concepts are at play. It is in such situations that the immensity and emergent possibilities of mathematics for teaching can come into focus.

Our purpose in engaging with the question of the meaning of 5/0 was not to home in on a correct response. Enhanced understanding of established mathematics was important but not the primary consideration. Our intention was to draw attention to the complexity of such constructs as 5/0, and in the process to grapple with the confusions and frustrations that young learners might experience. All of the teachers in the group confessed to having brushed aside "division by 0" in their teaching with the dismissive comment, "It cannot be done." In contrast, subsequent to this concept study, several of the teachers undertook studies of division by 0 with some of their classes, adapting discussions as appropriate to Grades 5, 7, 10, and 11. These teachers reported that not only were their students able to engage in nuanced discussions around diverse interpretations of number, division, and zero, they did so with an enthusiasm that was unusual for mathematics classes.

A Community of Experts

The timing of the 5/0 question was critical: it was posed after concept-study participants had engaged in their own self-directed inquiries for six months. The knowledge developed through their varied concept studies afforded the emergence of a "community of experts," in which carefully considered knowledge of number, division, zero and other constructs was woven into a grander productive conversation. We deliberately selected the question to highlight the manner in which expertise was distributed across the community and to interrupt a common, but troubling perspective on teaching – that each teacher should "know it all."

Because mathematics teachers are regularly positioned as subject-matter experts, there is a common expectation that every teacher must master *all* of the mathematical knowledge that he or she might encounter in class. The engagement around 5/0 interrupted this expectation by requiring varieties of expertise that inhered in the teacher collective as a whole but in no individual teacher within it. It offered an alternative model of knowledge wherein expertise is distributed in a community of practitioners.

The expectation that every teacher must "know it all" places considerable pressure on teachers, and can affect teachers' self-image. At least two elementary school teachers in this cohort told us at the start of the program that they knew "almost no mathematics." It was obvious that they were insecure about their mathematical knowledge and their ability to teach mathematics. Conversely, some high school teachers in the cohort saw themselves as "the experts" of the group, and felt compelled to offer answers when no one else could. Sometimes they appeared to be embarrassed when they did not have ready answers to questions that arose in our concept studies.

The "community of experts" model suggests a new organizational structure for knowledge in schools. It points to the myriad possibilities that can emerge when individual teachers come out of their classrooms to engage in a broader ongoing conversation about mathematics with their peers. It gestures toward a grander expertise that can only be enacted in collectivity. We believe that administrators and teachers should work together to create structures that enable collective conversations within their organizations. On an even grander scale, current digital technologies, such as digital forums and social media sites, allow conversations among experts to transcend the boundaries of individual institutions. One can imagine that, when used effectively, these technologies may give rise to very broad communities of experts that unite large communities of practitioners.

Graphically, we represent pedagogical problem solving as a new layer (see Figure 5.1) in the emergent collection of concept-study emphases. Once again, the image of nested circles is intended to flag that each layer is dependent on prior layers, but that the movement from one to another, or across all, is never linear. In this layer, teachers dealt with mathematical questions that arise in pedagogical practice, and in the process posed and addressed these questions:

- How would students interpret these mathematical questions at the different levels/grades?
- How are these mathematical questions similar to each other? How are they different?

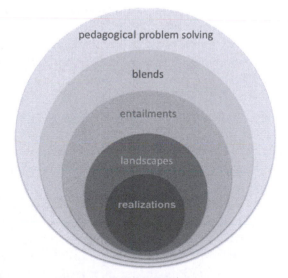

Figure 5.1. An emergent activity added to the earlier four concept-study emphases

- What difficulties would you expect students to have with these mathematical questions?
- What logical mistakes might occur? What analogical links and mis-links might be made?
- How could a teacher help bridge these transitions in understanding?

The new layer offers another contrast between the conception of M_4T developed in this book and more common understandings of teachers' disciplinary knowledge of mathematics within the research literature. As we see it, most research reports appear to be focused on the knowledge that teachers enact in classrooms situations. In terms of our five emphases, these reports are focused on the outer layer of pedagogical problem solving. Consider, for example, some of the entries on Ball and Bass's (2003) list of mathematical knowledge entailed in/by the work of teaching:

- selecting/designing tasks;
- identifying and working toward the mathematical goal of the lesson;
- listening to and interpreting students' responses;
- analyzing student work;
- teaching what counts as "mathematics" and "mathematical practice";
- making "error" a fruitful site for mathematical work;
- attending to ambiguity of specific words;
- deciding what to clarify, what to make more precise, what to leave in students' own language.

It is interesting for us to note the ways they these entries echo the questions that our concept-study participants posed for themselves while engaging in pedagogical problem solving (as noted in the previous paragraph). While we would not conflate these sorts of in-class teaching activities with the teacher-collaborative activities of pedagogical problem solving, it is important to be attentive to their common elements. Both are about establishing a complex coherence among the elements that might come together into a conceptual understanding.

In highlighting this similarity, we would also like to flag what we regard as two contrasting imbalances in the current literature. At one end, a prominent feature of much of the current resources for mathematics teacher education is the listing of interpretations of various foundational concepts (see, e.g., Van de Walle, Karp, & Bay-Williams, 2012). For the most part, these lists are offered without examination or nuance, in very much the same manner that teachers engage with varied realizations at the start of a concept study. At the other end, a great deal of the research into teachers' disciplinary knowledge of mathematics is concerned with the sorts of topics and questions that concept-study teachers grappled with during the emphasis on pedagogical problem solving (as noted a few paragraphs ago). In both cases, the vital and demanding interpretive work that connects the parts (realizations) to the whole (pedagogical problem solving) seem to be underappreciated. We worry that the the skills that enable abilities to substruct (represented in the intermediary circles of Figure 5.1) may be eclipsed by the more readily observable capacities to identify interpretations of concepts and/or to make appropriate decisions while teaching.

Of course, we in no way mean to suggest that educating teachers on diverse interpretations or focusing research on complex teaching decisions are inappropriate. Clearly, such knowledge and activities comprise large portions of the expertise of mathematics teachers and must be investigated. However, such foci might be obscuring the true depths of teachers' disciplinary knowledge of mathematics. As a consequence, we are troubled by trends toward formal evaluations of teachers' knowledge, since a research base sufficient to demonstrate the utility of such evaluations is yet to be developed. We suspect that until tools and methods are established that allow researchers to delve into the tacit and distributed dimensions of teachers' knowledge, most evaluation efforts will continue to fall short of securing the evidence base needed.

To this end, we contend that a participatory concept-study approach offers both a ready site and a ready means to investigate the tacit and distributed elements of teachers' knowledge – at the same time that it presents opportunities to nurture and evolve them. The emphasis of pedagogical problem solving relies on and extends the activities of acquainting, excavating, remembering, unpacking, blending – in a word, substructing – entailed in the other emphases.

Mathematics Teaching and Listening

In a recent concept-study session (with a Calgary-based cohort) devoted to pedagogical problem solving, a teacher brought up a question that was earlier asked by one of her students:

> When we solve algebraic equations, why do we eliminate fractional coefficients by multiplying by their reciprocals? In other words, why do we solve the equation $\frac{2}{5}x = 10$ by multiplying both its sides by $\frac{5}{2}$?

The teacher who reported this pedagogical problem noted that the student who posed it was able to get rid of the coefficient $\frac{2}{5}$ by multiplying both sides of the equation by 5 and then dividing them by 2. "He just doesn't see that multiplying by the reciprocal is the same thing," she said.

In true concept study fashion, the ensuing discussion focused on the realizations of multiplication, division, and fractions that the student might be working with. Even though the student was not present to corroborate the participants' speculations, we realized that the engagement was an exercise in orienting teachers to listen to the sense their students are making – not in order to correct, but rather with an intention to elaborate and blend current understandings into new frameworks with broader interpretive reach.

Earlier in this chapter we argued that pedagogical problem solving is a site of the real mathematical work of teachers. Solving pedagogical problems demands mathematical thinking that grapples with extant realizations, and considers their possible blends and entailments. Teachers who engage in this work must be closely attuned to their students' reasoning, as expressed through their verbal utterances and body language. In other words, teachers who enact M_4T as an open disposition must listen in a very specific way – in a manner that seeks variation, that participates with others in meaning making, and that is oriented to possibilities that may not yet be anticipated.

Our interest in teachers' modes of listening extends back much further than the decade of concept studies described in this book (see Davis, 1994, 1996, 1997). Still, it was not until we engaged with the pedagogical problem solving emphasis that we began to truly appreciate the profound relationship between how mathematics teachers listen to their students and how they know their mathematics. Before we examine this relationship more closely, we shall first summarize the three "modes of listening" that teachers engage with their students and their mathematics – evaluative, interpretative, and hermeneutic – as identified in Brent's earlier work (e.g., Davis, 1996). These modes were identified on the basis of classroom observations, taped teacher-student interactions, and the resulting transcripts. These data were all used to track significant shifts in learner understanding (or lack thereof), and to link these shifts back to the qualities of interaction with the teacher.

Easily the most common mode of listening observed across many contexts was *evaluative listening*. In this mode, the teacher would typically be listening for specific responses, providing immediate feedback, most commonly in the form of a statement on the correctness of student contributions. In evaluative listening, there is rarely any significant wait time prior to the response, and it is even rarer for the teacher to pause after or linger on an answer. Rather, the teacher appears to be using responses by the student as expected markers along a pre-selected conversational trajectory.

At first hearing, this description might give the impression that evaluative listening is necessarily cold and detached, but this need not be the case. Every teacher in the study employed evaluative listening on some occasions. The fact of the matter is that many of the interactions that mathematics teachers have with their students are focused on ensuring that students have adequate mastery of standard curriculum. With some teachers, however, it seemed that evaluative listening was their *only* mode of listening. Students of these teachers, perhaps not surprisingly, were less able to talk meaningfully about their mathematical understandings (Davis, 1989).

Interpretive listening is rooted in the conviction that learners' interpretations are never inconsequential or random. The teacher therefore has an obligation to make sense of where the student is coming from. Interpretive listening can manifest in simple teaching acts of pausing to ask a respondent to rethink an answer or to explain how a particular conclusion was drawn. In the study, these acts have been associated with significant differences in students' abilities and their willingness to talk about their mathematics.

An ongoing quest for diverse responses is a hallmark of interpretive listening in whole-class situations. Teachers in the study would often ask students to give incorrect answers, along with explanations, in order to highlight where and how someone might go wrong with a concept or procedure. A common accompanying question, for example, was, "Why might someone think that?" This sort of query invited students to appreciate that errors and misconceptions can be valuable sites of learning.

The classrooms of teachers who listen interpretively were found to be considerably more interactive than those of teachers who listen only evaluatively. There were many more student articulations, and these tended to be longer and more sophisticated. Students also tended to ask questions more freely, to seek clarification, and occasionally to question the teacher's explanations. However, one feature was common to both the evaluative and interpretive classes: the teacher was the authority on mathematics. While there was greater openness to the idiosyncrasies of individual sense-making in the interpretive listener's classroom, the sense that there exists a right way to answer or a best way to explain persisted.

Hermeneutic listening is a mode of engagement in which the teacher deliberately seeks difference of opinion and encourages variety of thought. The goal is *not* to find the best answer, but to infuse the collective with a richness of interpretation. This mode of listening is guided by the assumptions that there are always new ways to talk about mathematical ideas and that it is always possible to improve on understandings, explanations, and interpretations. In the classes of teachers who employ hermeneutic listening, the locus of learning seems to shift outside of the individual and into the space of the collective. The social structure of the classroom also seems to shift, modulating organically between centralized (where a student or the teacher holds the floor for an extended period) to decentralized (where discussion continues in multiple small groups that influence one another).

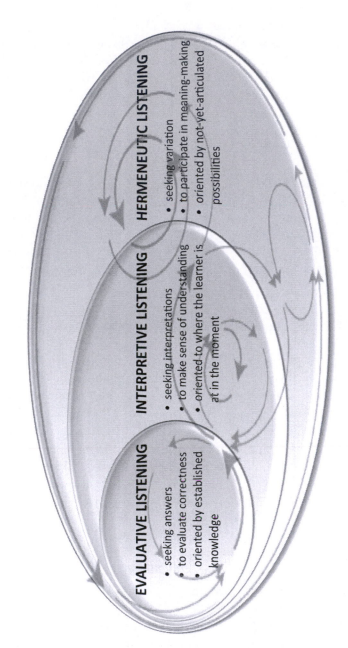

Figure 5.2. Three co-implicated modes of listening

To contrast the three modes of teaching-as-listening in a sentence: the lesson trajectory of the evaluative listener-teacher is largely *unaffected* by student articulations, the lesson trajectory of the interpretative listener-teacher is *modified* by them, and the lesson trajectory of the hermeneutic listener-teacher is *defined* by them. Alternatively, in temporal terms, evaluative listening focuses on the past with its concern for fidelity to already-established knowledge; interpretive listening adds a concern with the present as it attends to learners' unfolding interpretations; and hermeneutic listening includes and transcends the others by layering an interest in future, emergent possibilities. Graphically, one might portray evaluative, interpretive, and hermeneutic listening as three nested circles, where each outer circle presupposes but transcends the ones within it (see Figure 5.2).

Returning to the relationship between teachers' disciplinary knowledge of mathematics and their listening attitudes, it was only when concept study participants engaged in pedagogical problem solving that it dawned on us that the two are inextricably intertwined. M_4T, as we understand it, is an open disposition toward the collective production of new insights. It is dependent on and enabled by one's formal knowledge of mathematics, but is in no way reducible to it. The transition from knowledge of formal mathematics to knowledge of M_4T entails an open, participatory mode of attendance. This mode, we believe, is illustrated in all of the anecdotes that open chapters in this book, and will be further illustrated by the teaching examples in Chapter 6.

For us, the past decade of concept study research into teachers' disciplinary knowledge of mathematics has helped clarify two important requisites of hermeneutic listening. The first is a substantial depth and breadth of mathematical knowledge. In order to listen hermeneutically, teachers must possess a rich reserve of personal understandings, in order to prevent them from shutting down others' efforts at making sense of mathematical concepts. The second is a thorough appreciation of the vital connection between mathematical knowledge and the manner in which it is produced. In considering how mathematical knowledge is produced, issues of personal efficacy, social systems, and cultural values come into play. The hermeneutic listener will consider them to be an *integral part of* teachers' knowledge of mathematics, and not merely a means to enabling student learning of formal mathematics. We explore the two requisites for hermeneutic listening in more detail in the next two sections.

A Depth and Breadth of Mathematical Knowledge

Up to this point, we have been focusing mainly on the necessary pedagogical depth of teachers' knowledge of mathematics. A matter that might have been given short shrift is the breadth or extant of teachers' mathematics knowledge

— and, in particular, the relevance of advanced studies in mathematics (which so occupied the attentions of Begle and other researchers in early studies of disciplinary knowledge for teaching). The pedagogical problem solving emphasis helped to highlight our own narrowness in this regard.

As already mentioned, as part of their work in the master's cohort, participants took responsibility for structuring their own concept studies. These were conducted outside of class times for the most part. To assist in these studies and to monitor progress, additional whole-group sessions were scheduled. In each of these two to three-hour meetings, one group of participants led the activities, and only one or two mathematical concepts were addressed.

In the first few sessions, the participants focused on pedagogical depth. They were guided by and adhered to the four emphases introduced in the research methods course. In other words, they set out to list realizations, develop landscapes, consider analogical entailments, and combine realizations into blends. Sometimes this approach worked well, but sometimes results were scant, and did not reflect the considerable efforts put forward.

It did not take long to figure out why this was the case: abstract mathematical concepts, drawn from high school algebra, do not lend themselves to the four emphases as readily as do more concrete concepts, drawn from elementary school arithmetic. There is a dramatic shift in abstraction and complexity as one moves from arithmetic to algebra. Many concepts of arithmetic can be readily interpreted in terms of everyday, immediate experience, and so the activities of identifying realizations, formatting them into meaningful landscapes, and exploring some of their entailments are reasonably straightforward. In contrast, many algebraic concepts do not have ready realizations in everyday experience. Physical representations of algebraic concepts are not everyday objects, but abstractions themselves, as evidenced by the algebraic manipulatives associated with school algebra (e.g., algebra tiles, two-pan scales). To put it differently, whereas the objects of school arithmetic can be regarded as abstractions of everyday experience, the objects of school algebra are abstractions of mathematical objects and processes. It is not surprising that substructing algebraic concepts involves added layers of abstraction and complexity, and perhaps requires different concept study emphases.

This point was brought home during a concept study on factoring. The small group of high school teachers who selected the topic began the session by asking participants to think about realizations they might have encountered. Unlike our experience with arithmetic concepts (e.g., multiplication, zero) where long lists of realizations were generated readily, the few interpretations of factoring that came forward were the ones commonly found in textbooks (e.g, manufactured manipulatives). Moreover, only the high school teachers were able to identify these realizations. There was no point in pursuing landscapes, entailments, and blends for this concept.[3]

The difficulty in adhering to the four emphases was not in itself problematic. After all, the point of concept study is not to impose a fixed set of procedures in order to generate a uniform product. It is, rather, to come to a deeper, more explicit awareness of the subtle complexities of a mathematical concept in order to teach it better. More troubling was the frustration that the high school teachers experienced in attempting to devise alternative means for substructing algebraic concepts. The difficulties they experienced in generating multiple realizations for algebraic concepts indicated that they were not engaging their students in anything deeper than unitary, procedure-based understandings of these concepts.

The concept studies that addressed high school mathematical concepts revealed a need for mathematical breadth – that is, teachers can greatly benefit from knowledge of mathematics that extends beyond what is expected of their students. This finding supplements Baumert and colleagues' (2010) assertion, mentioned in Chapter 1, that content knowledge of mathematics "remains inert in the classroom unless accompanied by a rich repertoire of mathematical knowledge and skills relating directly to the curriculum, instruction, and student learning" (p. 139). We contend that a forward-looking awareness of more advanced mathematics that the students are likely to encounter in later years is a vital component of any teacher's repertoire.

Much work remains to be done before any strong claims can be made, but based on our decade of experience with teachers substructing their mathematics, we feel that the time has come for some preliminary speculation on topics in advanced mathematics that might be particularly useful to teachers of elementary mathematics. The topics that came up repeatedly in concept studies included partitioning, continuity, and $0/0$. They suggest that current requirements for coursework in number theory and calculus might be defensible.

As important as these areas of advanced mathematics might be, the non-substructive manner in which they are often taught in university courses might render their study less useful than is desired. Moreover, recall that in a community of experts, it is not important that everyone present is

---

3. Just for the sake of interest, the factoring group had some success in the activity of tracking the emergence of concepts that contribute to the process of factoring throughout the curriculum. One surprising result that arose in this group's study was that, within high school algebra, the inverse/opposite of factoring is not a simple product, but rather a distributed product of multiplication over addition. This realization prompted some lively discussions among teachers across grade levels on how they might support more sophisticated understandings in earlier grades.

knowledgeable in these areas. Since it suffices that the requisite knowledge be part of the knowledge-producing system, one may argue that teachers should study a broader array of advanced mathematical topics than is currently mandated in most teacher preparation settings. We believe that a combination of greater diversity in mathematics backgrounds *alongside* formal, collective structures to substruct concepts will greatly advance the vibrant body of teachers' disciplinary knowledge of mathematics, that is, M_4T.

Knowledge of How Mathematics is Produced

One of the current trends in the research into mathematical knowledge for teaching is to identify sub-categories of such knowledge. Perhaps the best known of these is drawn from the extensive research of Deborah Ball and Hyman Bass through their Learning Mathematics for Teaching project. Ball and Bass[4] make a number of fine-grained distinctions in their analysis of knowledge of mathematics for teaching (KMT), starting with a distinction between Subject Matter Knowledge and Pedagogical Content Knowledge. Further subdistinctions are presented in Figure 5.3.

The distinctions offered by Ball and colleagues are somewhat analogous to ones identified in another analysis that was based on concept-study research (Davis & Simmt, 2006; see Figure 5.4). In particular, we see our *categories of knowledge* as paralleling their *subject matter knowledge* and our *categories of knowing* as paralleling their *pedagogical content knowledge*.

We also perceive an important conceptual difference between the two models. It is signaled by our distinguishing of *knowledge* from *knowing*. As we noted earlier, in the context of mathematics teaching, knowledge and knowing are inseparable, co-implicated phenomena. In complexity research terms, the differences between what is usually called "knowledge" and what might be described as "knowing" are mainly matters of scale and timeframe. Mathematics knowledge on the collective level evolves at a pace and a scale that can make it seem fixed when viewed from within the frames normally inhabited by individuals.[5] Conversely, mathematical knowing on the personal level can self-transform so rapidly that it appears volatile and unstable against collective knowledge.

---

4. Visit http://sitemaker.umich.edu/lmt/home for descriptions, articles, and supporting materials. See also Ball, Thames, & Phelps (2008).

5. The point being made here might be aligned with Shulman's (1986) distinction between "the substantive and the syntactic structures" of a discipline. As mentioned in Chapter 1, he noted, the "substantive structures are the variety of ways in which the basic concepts and principles of the discipline are organized to incorporate its facts. The syntactic structure of a discipline is the set of ways in which truth or falsehood, validity or invalidity, are established" (p. 9). In brief, we agree that these are vital aspects of pedagogical content knowledge.

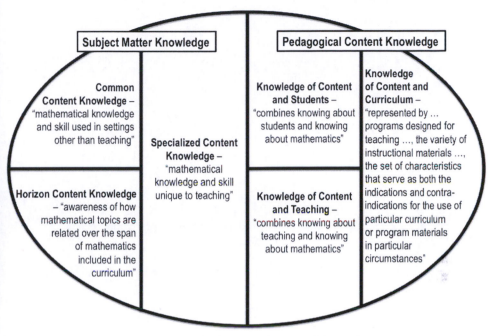

Subject Matter Knowledge

Common Content Knowledge – "mathematical knowledge and skill used in settings other than teaching"

Specialized Content Knowledge – "mathematical knowledge and skill unique to teaching"

Horizon Content Knowledge – "awareness of how mathematical topics are related over the span of mathematics included in the curriculum"

Pedagogical Content Knowledge

Knowledge of Content and Students – "combines knowing about students and knowing about mathematics"

Knowledge of Content and Teaching – "combines knowing about teaching and knowing about mathematics"

Knowledge of Content and Curriculum – "represented by … programs designed for teaching …, the variety of instructional materials …, the set of characteristics that serve as both the indications and contra-indications for the use of particular curriculum or program materials in particular circumstances"

Figure 5.3. Ball and colleague's framework for analyzing Knowledge of Mathematics for Teaching (KMT). (Image and descriptions adapted from Ball, Thames, & Phelps, 2008)

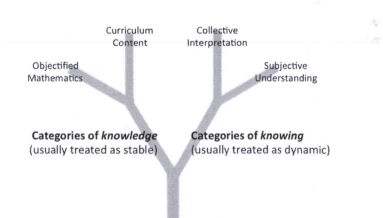

Curriculum Content

Collective Interpretation

Objectified Mathematics

Subjective Understanding

Categories of *knowledge* (usually treated as stable)

Categories of *knowing* (usually treated as dynamic)

MATHEMATICS-FOR-TEACHING

Figure 5.4. Davis and Simmt's (2006) aspects of mathematics-for-teaching (M_4T)

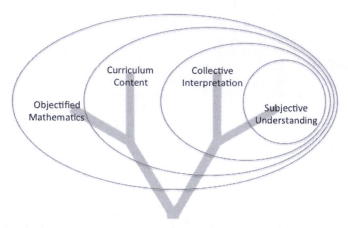

Figure 5.5. Interpreting knowing and knowledge as nested phenomena

These categories, then, are perhaps better portrayed as nested phenomena than as neighboring regions. Visually, we might illustrate the co-implicated phenomena by overlaying a set of nested circles on the M_4T tree (see Figure 5.5). The resulting image highlights the complex manners in which different levels of mathematical process/product unfold from and are enfolded in one another. Our focus, within this complex framework, is on the myriad ways in which humans engage with mathematics. In particular we are interested in the individual, social, institutional, and cultural dimensions of the generation of mathematical meanings.

This same focus is at the heart of concept study and, as noted above, we believe it to be a core element of teachers' disciplinary knowledge of mathematics. From a pedagogical standpoint, one cannot separate one's knowing/knowledge of a discipline from one's knowing/knowledge of how that discipline is learned. Concept studies are deliberately structured to foreground both elements simultaneously.

At any given concept study moment, participants modulate between references to mathematics knowledge (i.e., left side of the image) and mathematics learning (i.e., right side of the image). This classroom anecdote, provided by one of the cohort's participants, illustrates the duality of knowledge/knowing.

[VOICE: LENA]

Last month, my Grade 3 class [age ~9 years] was counting down from 95 by threes. When they got to 2, I thought they would stop. But the counting continued to −1, −4 and so on. After a while, they stopped and realized that the numbers were increasing in magnitude. Then a boy deep in thought blurted out a question, "Can zero be a negative number?"

The whole class became alive and engaged in hypothesis, proof, different representations on the chart paper, theorizing, and a genuine pursuit of knowledge. Oh, the joy of zero. I could have answered 'no' but we would have missed a very rich learning moment. It was a transformative moment for that boy and for the rest of the class. It opened up the possibility of thinking of math in different ways. The experience definitely transformed my thinking about counting backwards, zero, and the kind of mathematics you need.

To re-emphasize, the point we are making in this section is that, for teachers, it is important to be aware that the generation of personal understanding is nested in grander systems of knowledge production. This brief narrative is about more than a group of children coming to personal understandings of a pre-given, unchanging set of facts. It is also about children's getting acquainted with the dynamics of knowledge production, and about their contributing to the conceptual richness of zero in ways that expand personal and shared horizons.

Where Are We Going with This?

One of the most encouraging results in our two years of working with the master's cohort described in Chapters 4 and 5 was the manner in which participants took ownership of the research methodology of concept study. While all attempted to employ the four emphases – realizations, landscapes, blends, and entailments – in their own concept study projects, no one adhered to them "religiously." Rather, the emphases were treated as strategies that may or may not be useful, depending on the concept under study.

As we recounted in this chapter, we were initially taken aback when we discovered how unhelpful the four emphases turned out to be when applied to some more abstract high school concepts. In retrospect, we should not have been so surprised. From the beginning of our work with concept study, participating elementary-school teachers have tended to adopt and adapt its emphases for their classrooms. Secondary school teachers, on the other hand, did not do that. Lena's account about zero illustrates this point. We received it when Lena was only a few months into the program. Yet, we have no similar anecdotes from the secondary school teachers in the program dating to that period.

This contrast is all the more interesting, given the more extensive disciplinary backgrounds of the high school teachers in the concept study group. Many of these teachers arrived at the master's program with far more confidence in their knowledge than did their elementary counterparts. As we explained earlier, we believe that the problem lies with the profound difference in conceptual complexity between elementary and high school concepts. Acknowledging this explosive increase in complexity is essential if we are to understand the difficulty of "teaching for understanding" in the higher grades.

Even though such teaching is difficult, as many teachers have demonstrated, it is far from impossible. In Chapter 6, we visit the classroom of one high school teacher who brilliantly used concept study sensibilities in his teaching.

CONCEPT STUDY IN THE CLASSROOM

ENACTING AN OPEN WAY OF BEING

— In brief …

This chapter recounts the first deliberate application of concept study sensibilities in a mathematics classroom. In the process, we have encountered a world of possibilities that open up when teachers enact an open disposition that is receptive to emergent mathematics in its many forms.

A Concept Study of Circles

[VOICES: BRENT AND MOSHE]

In the fall term of 2011, cohort participants were busily working on their culminating concept study projects. Many had organized into groups in order to study topics such as zero, equality, and fractions.

Quite unusual among them was Freddie, an experienced mathematics teacher and department head, who chose the topic of circles. Rather than working with cohort colleagues, however, he approached us and asked for permission to conduct his study with some of his students.

We were somewhat skeptical at first. All concept studies up to this point focused on concepts of arithmetic and, to a lesser extent, on algebra. It was not easy for us to imagine the network of associations surrounding circles, and we wondered if the concept-study emphases developed to that point (i.e., realizations, landscapes, entailments, and blends) would apply to circles at all. Moreover, our concept studies to date had been conducted with groups of practicing teachers and sought to explore the collective mathematical knowledge that inheres in such groups. What collective knowledge about circles could possibly inhere in a group of 8th-grade students?

Despite our initial skepticism, we were also intrigued. Freddie was an expansive and creative thinker in the cohort. In earlier conversations, he had made it clear that, for him, teaching with an open disposition towards classroom emergence is the only way to teach. If anyone could pull off a concept study on circles, it would be him. And so, we offered to participate actively in the study in order to have a first-hand experience of it.

The illustrative narrative in this chapter is structured somewhat differently from those of the preceding chapters. It unfolds gradually through the chapter, affording us opportunities both to highlight and to work with its unfolding complexity. Our intention is to offer a glimpse of the emergent possibilities that a concept study gives rise to in a mathematics classroom wherein the teacher is deeply committed to students' development of mathematical meanings, and not only to their mastery of mathematical definitions and procedures.

It is important to note that this narrative was drawn from a secondary school (grade 8) classroom. In our experience, many teaching innovations at this level are brushed aside with the claim that, while pedagogical innovation might be possible in elementary school, teachers of higher grades are far too busy, as they need to focus on curriculum coverage for upcoming exams. But as the teaching events in this chapter illustrate, the imperatives of testing are poor substitutes for the possibilities that emerge in a meaning-laden and collectively enabled high school mathematics classroom.

In terms of the five concept-study emphases that have been developed so far, we see the examples provided in this chapter as elaborations of pedagogical problem solving. In other words, the teachers' disciplinary knowledge of mathematics is activated and articulated as they enact an open disposition in responding to and working with the mathematical questions that animate students' attentions.

Four Quadrants of M_4T

Freddie was drawn to Wilber's (1995, 2006) integral theory, introduced in our research methods course as a means to foreground the similarities and disconnects among varied research traditions. Integral theory views reality as co-arising and co-manifesting in four co-implicated realms: objective, subjective, intersubjective, and interobjective. Freddie used integral theory to analyze the events that transpired in the classroom. Likewise, we use Wilber's model here to conceptualize the concept study on circles.

Developed by the American philosopher Ken Wilber over the past three decades, integral theory is a large-scale framework for making connections among diverse theories of social reality. It offers an analytical space for the systematic integration of multiple knowledge traditions and paradigms.

Fundamental to the integral model is the notion of *four quadrants*. Wilber (1995) suggested that the investigation of any psychosocial phenomenon must take two fundamental perspectives into account. The interior-exterior

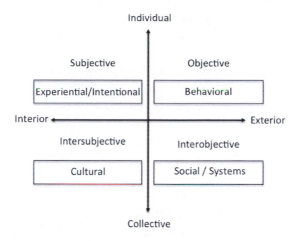

Figure 6.1. The four quadrants of Integral Theory

perspective refers to the relationship between subjective experience and objective behavior. The individual-collective perspective refers to the relationship between the personal and the social. The two perspectives combine (see Figure 6.1) to yield the four quadrants – experiential (subjective), behavioral (objective), cultural (intersubjective), and social (interobjective).

The quadrants are four interrelated domains of reality and also four perspectives through which we can gain access to these domains. For example, when we feel elated after listening to a performance of a violin concerto, the experience of elation can be understood in different ways. From a subjective perspective, we experienced a transcendent feeling of transformation that made us very excited. From an objective perspective, sound waves vibrated in our ear and caused specific neural activity in parts of the brain. From an intersubjective perspective, our culture attaches emotional value to the activity of listening to music. From an interobjective perspective, the piece we listened to belonged to the canon of Western music, which is a specific system for organizing sound.

In other writings (Renert & Davis, 2010a; Renert & Davis, 2010b), we surveyed M_4T through the lens of the quadrants, and offered four fundamental dimensions of M_4T (Figure 6.2, upper). The objective dimension deals primarily with the objects of mathematics – fractions, Pythagoras' theorem, and the like. The subjective (interior-singular) dimension deals with personal meanings, emotions, and attitudes associated with the teaching of mathematics. The intersubjective (interior-plural) dimension deals with shared meanings and values. And the interobjective (exterior-plural) dimension deals with external systems that enfold and are enfolded in mathematics and its teaching, such as systems of schooling. In the act of teaching, these foci correspond to

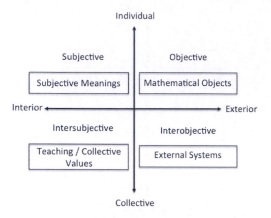

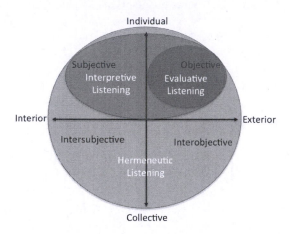

Figure 6.2. Quadrants of M₄T (upper) and listening (lower)

different modes of teacher listening, as introduced in Chapter 5 (Figure 6.2, lower). A focus on objective knowledge aligns with an evaluative listening; a broadened interest in subjective interpretation compels and interpretive listening; awarenesses of intersubjective and interobjective dimensions of knowing compels a hermeneutic listening. (A similar correspondence can be drawn between the fours aspects of M₄T, illustrated in Figure 5.4, and the four quadrants of M₄T, illustrated at the top of this page.)

The four quadrants represent four irreducible domains. Awareness of the four quadrants and their underlying dynamics can greatly broaden a teacher's field of vision. A common reductionist mistake, which Wilber called "quadrant absolutism," is to privilege one quadrant while devaluing (or even excluding) the others. Returning to the above example of a violin concerto, a scientific description of the event that focuses solely on brain activity necessarily misses out on much of the vitality of the experience. To be sure, the feeling of elation in the experience of music has correlates in all quadrants. But the most exciting part of the experience probably resides in the subjective quadrant of intangible experience.

Early on, Freddie noted that, in at least one respect, our framing of concept study could be accused of quadrant absolutism. He pointed out that the opening question we have commonly asked, "What is X?" (where X is the mathematical object under study, e.g., *What is multiplication?*), suggests that the goal of concept study is to *define* the mathematical object. The search for a definition, in turn, might lead participants in concept study to privilege the objective dimensions of X over all other dimensions. With this in mind, we settled on an alternative opening question – "What is interesting about X?" – as a possible means to activating also the subjective, intersubjective, and interobjective dimensions of the concept under study.

What's Interesting about Circles? Emergent Problem Solving

[VOICES: MOSHE AND FREDDIE]

Our study of circles began with a video that we found by searching for the keyword "circles" on YouTube. The theme was circles found in nature. Images of circular galaxies changed into images of volcanic smoke rings, which in turn transformed into images of dolphins playing with ring bubbles. We were all fascinated by the beauty of the images, and the students asked to watch the video again. At the end of the viewing, we sensed a feeling of great expansiveness in the room.

We proceeded to study and derive some of the more standard mathematics of circles: the number π, and formulas for circumference and area of a circle. We then told the students that they would decide what we should study about circles in the next few lessons. We asked them to get together into groups and respond to the question, "What's interesting about circles?"

After about 20 minutes of group consideration, the students came up with many questions. Through discussion, the class settled on two of them:

1) How many sides does a circle have? One or infinitely many?

2) Why is the number π so mysterious?

We took a vote and over four-fifths of the class opted for the first question.

> *Teacher: Why do you think that a circle has one side?*
> *Student: It just does.*

The student traced an imaginary circumference with her finger.

> *Teacher: So why does a circle have infinitely many sides?*

Two students came up to the blackboard and drew a sequence of geometric shapes (Figure 6.3).

Figure 6.3. A sequence of polygons with increasing number of sides

> *Student: As the number of sides increases, the shape looks more and more like a circle. So when there are infinitely many sides, it must be a circle.*
>
> *Teacher: Your drawing shows figures with three or more sides, starting with the triangle. And we said that a circle might have only one side. So what about figures with two sides? Can you think of some?*

The students paused to think. Two suggestions were offered: a semi-circle and a two-sided figure drawn on a sphere (Figure 6.4).

> *Teacher: Are we allowing our figures to be drawn on spheres now, or are we limiting them to the two-dimensional Euclidean plane?*

A lively debate then began, but the bell rang and we had to stop for the day. We were off to explore the question, How many sides does a circle have? This was not a question we had ever contemplated. So we were all in it together. This question would occupy us for the next three class sessions.

Problem solving has been a major focus of mathematics pedagogy since the late 1970s (e.g., Schoenfeld, 1985). In particular, much has been written and researched about the importance of open-ended questions and learner-posed problems. Despite the extensive literature, it seems, in our experience at least, that little has changed around mathematical problem solving in classrooms over the past few decades. Most "problems" continue to be drawn from textbooks and other published resources and most "solving" continues to rely on the unchallenged assumptions that each problem has one correct answer and that the teacher (or textbook author) knows this answer. In other

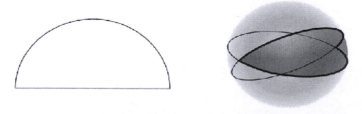

Figure 6.4. Polygons with two sides

words, students' creativity is limited to replicating solutions that are already known by an adult.

But the concept study sensibility of "What's interesting about circles?" appears to have challenged these longstanding assumptions. The two problems posed by the students in this study – *How many sides does a circle have?* and *Why is the number π so mysterious?* – gesture towards a different kind of problem solving. For one, their solutions were not known *a priori* by us, the teachers. Indeed, we were not certain that satisfying answers could be found at all.

With respect to the latter question about π, we were delighted that it drew so clearly from the subjective and intersubjective quadrants of circles. The adjective "mysterious" indicates both personal fascination and collective judgment. We could imagine future engagements with this question proceeding not only into explorations of properties of π and characteristics of irrational numbers, but also into evaluations of what constitutes a "mysterious" number as opposed to an "obvious" number.

Likewise, another world of exciting investigations opened up when the students chose the question about the number of sides of a circle. The sequence of polygons with increasing number of sides (Figure 6.3) clearly pointed to the concept of *limit*, and the polygon drawn on the sphere (Figure 6.4) pointed to *non-Euclidean geometry*. Both of these mathematical topics are normally considered to be the purview of higher mathematics, inappropriate for 8th-grade students. And yet both of them manifested in the collective knowledge of the grade-8 group in response to the open-ended question *"what's interesting about circles?"* We argue that both concepts already reside tacitly in the collective's knowledge base, ready to be explicated by the structures of concept study.

The new type of problem solving also forced us to reconsider our roles as teachers. Since the questions were emergent, and did not resemble standard textbook questions with which we were familiar, we did not have clear answers to them. And so, we were no longer unassailable knowledge authorities. Instead, we became co-participants in the investigation. We were members of the knowledge-producing collective seeking to increase our understanding, just as did other members. Admittedly, due to our more extensive mathematical histories, we would probably be able to contribute quite richly to the investigation. Yet, we suspected that a different order of ingenuity would be required to approach these emergent problems. From a complexity perspective, we were experiencing the transformation of learning to novel and more complex patterns of organization.

Do Circles Exist? Student Agency in Mathematical Knowledge Production

[VOICES: MOSHE AND FREDDIE]

Our investigations into the question of how many sides a circle has took many unexpected turns. Quite early on, we discovered the need to be very precise about our terminology and the ways in which we use terms. What is meant by "side"? Must a side be straight? If not, in what sense does a triangle have three sides?

When the students researched these questions on the Internet, they came across some different, and occasionally conflicting, answers. Since the terms "polygon" and "vertices" were used in some of the answers, the class had to incorporate them into the discourse. They also came upon new and unusual mathematical figures: the digon, a two-sided polygon on a spherical space, and the apeirogon, an open polygon with infinitely many sides (Figure 6.5). All these discoveries brought up even more questions. Is a circle a polygon? What makes an octagon an octagon — its eight vertices, its eight sides, or both? Can a polygon cross itself? Does a polygon need to be closed?

In this context, one student brought an interesting perspective to bear.

Student: *There is no such thing as a circle.*
Teacher: *Why are you saying that?*
Student: *When you draw a circle with a pencil, the tip of the pencil has a certain thickness. Every time you move the tip, you create a very short side. So even though it looks like a circle, it is really a polygon with lots of sides.*

Another student responded:

Student: *But we know what a circle is. A circle is a collection of points, and points have no thickness. So, if we can think about a circle, it must exist, even though we cannot draw it accurately.*

The students looked puzzled.

Teacher: *What does it mean when we say that something "exists?" Does it need to exist physically or only in our minds?*

The students were looking at us, hoping that we would sort this out for them — one of the biggest questions of philosophy no less. What aspect of teacher preparation, we wondered, should have prepared us for this moment?

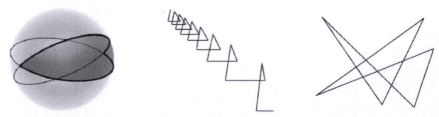

Figure 6.5. Unusual polygons: a digon, an apeirogon, and a polygon that crosses itself

As the preceding classroom episode illustrates, the question of the number of sides of a circle quickly boiled down to issues of meanings and definitions. In order to proceed with their investigations, the students needed definitions for some very rudimentary mathematical notions, such as *side*. As these definitions were not readily available in their textbook, the students opted to use the Internet in their research. In doing so, they were surprised to discover that mathematical terms have contested, sometimes conflicting, meanings.

As the concept study proceeded, the students became increasingly aware of the role that language plays in humans' construction of mathematics. We noticed that the ontological status of mathematics began to shift for the collective. No longer was mathematics just knowledge that is "out there," representing eternal truths that are divorced of their knowers. On the contrary, the more choices the students were making among competing definitions and understandings, the greater agency they felt in *creating mathematics*.

At one point, we wanted to consolidate the class' research on the question of how many sides a circle has. We asked the students to write down the answers that they had found and any remaining questions. There were two answers and over two dozen new questions. In the discussion that followed, the students argued that there cannot be a definitive answer to the question of the number of sides of a circle. Answers depended on meanings of terms such as "sides" and differed depending on whether the circles were embedded only in the Euclidean plane.

In order to be provocative, we proposed to invite a geometry specialist from the university to visit our class and to settle the matter. The students liked the idea of inviting an expert guest, but argued that his answer would just be one of many opinions. They were clearly comfortable with their constructed mathematics, and were willing to defend it, even in the presence of an expert.

Throughout the process, the students exhibited increasingly confident agency in the mathematics. Not only did they pose the original questions, but also they were the ones to propose solutions and then to analyze the validity of these solutions. Their activities involved much more than what we typically call mathematical problem solving in classrooms. The structures of knowledge production – for example, language, definition, conjecture, qualification – were readily available for examination and challenge. We noticed an ongoing interplay between engagement with the stated goal of "solving the problem," and the complementary meta-engagement in the analysis and critique of the mathematical structures used in the solution.

The range of mathematical knowledge required of us as teachers kept broadening as the study progressed For example, when the students came up with the examples of the polygons in Figure 6.5, we had to adjust our working definitions of *polygon* to suit. When asked if a line segment must be straight or could it curve or snake, we had to admit that we simply did not know. But when the philosophical question arose about *"what does it mean when something exists?"* we realized that a concept study engagement with students can, and is likely to, expand M_4T – that is, the mathematics that teachers need to know in order to teach mathematics – far beyond what we had found in the research literature so far.

Rope around the Equator: Cultural Mathematics

[VOICES: MOSHE AND FREDDIE]

The class meetings did not all focus on the question of the number of sides of a circle. Sometimes the investigations of this question ran into dead ends. We used these opportunities to consider other interesting aspects of circles. The students brought up some of them and, as co-participants in the study, we felt that it was appropriate for us to suggest topics that we found interesting too. The following interaction took place around a problem that has puzzled us for a while.

> *Suppose that the earth is a perfect sphere and that we tie a rope tightly around the equator. How long will that rope be? What information do you need to know in order to answer this question?*

The students asked how long the radius of the earth was, and we told them that it was 6391 kilometers long. Individually and in small groups, they quickly calculated that the rope would be about 40 000 kilometers long.

> *Now, suppose that we don't wish the rope to be so tight. I add 10 more meters of rope to the existing 40 000 kilometers, and then loosen the new longer rope evenly around the equator. Do you all agree that a gap will form between the earth and the rope?*

The students nodded in agreement.

> *How tall will this gap be? What could pass through it? An amoeba, an ant, a child, or a mountain? Please give us your best estimate.*

We took a vote. Almost all of the students thought that an ant could pass through the gap. A few thought that the gap would be too small for an ant to pass through, and voted for the amoeba.

> *Okay. Now please calculate the height of the gap.*

The students worked out some answers. After some discussion, everyone agreed that the correct answer was about 1.6 meters long. In other words, the gap is tall enough to let a child pass through. The students appeared very surprised. We then told them that this result had surprised us too when we first came across it, and that it still does. Intuitively, it seems that when we lengthen a rope of 40 000 kilometers by 10 meters, the loosening effect should be negligible.

How can it be that the gap is large enough to allow a child through? How would you explain this result to a person who cannot calculate it?

We posed this question to the students, even though we did not know the answer ourselves. The students took about 15 minutes to come up with answers. A number of explanations were offered but they were mostly verbal reiterations of the calculations that were just performed. Then one of the students raised his hand and provided this answer:

For us, a gap of 1.6 meters looks big. But this gap 1.6 meters is added to the radius of the earth. If you compare 1.6 meters to the radius of the earth, which is 6391 kilometers, you can see that it is not large at all. In fact, it's tiny.

There it was – a perfectly clear and sensible explanation that had eluded us for years. The proportional increase of 1.6 meters is negligible, even though a child can pass through the gap. The student was able to explain the result even though common intuition was unreliable in thinking through the question.

The student's explanation was not one that we would expect to find in a mathematics textbook. Yet new mathematics was created right there, in the moment, and it was brilliant.[1] What kind of mathematics is it? We call it *cultural mathematics*. We use the term to refer to any and all mathematics that extends *formal mathematics*. To be precise, we view formal mathematics as mathematicians' mathematics – the canon of mathematical results, developed over thousands of years, which is neatly summarized in school mathematics textbooks. Cultural mathematics is all that resides outside of that. It is the analogies, metaphors, applications, systems, discourses, and practices that relate to mathematics but are not traditionally seen as formal mathematics.

Much of school mathematics revolves around the codified results of formal mathematics. A common perception of teachers, students, and parents alike is that mathematics education is about the transmission of these results from one generation to the next. This is why students are often graded on their faithful replication of pre-established results. In reality though, most current research in mathematics education focuses on the need to teach not *mathematics* but rather *the mathematical*, that is, the processes, dispositions, and habits of mind required to engage with the mathematics in all its guises – both

---

1. The nature of this insight and its mathematical significance are more apparent if one attempts the problem arithmetically rather than algebraically. The precision required to determine the difference is beyond that offered by most calculators.

formal and cultural. Put differently, when considering the four quadrants of M_4T (Figure 6.2), the results of formal mathematics reside almost exclusively in the upper-right objective quadrant, while cultural mathematics accounts for all four quadrants. For the sake of illustration, cultural mathematics includes the belief that "math is hard" and that people who are good at it are "geniuses," SOHCAHTOA, counting on one's fingers, poker games, a career in engineering, sales taxes, daily quizzes, a preference of degrees to radians, and a rope around the equator.

Cultural mathematics forms much of the hidden curriculum of our mathematics classes. Sometimes, teachers treat it as a nuisance or "noise" (e.g., Jimmy was counting on his fingers but he should really stop doing that). At other times it is treated as a crutch at the service of formal mathematics (e.g., I teach my students the mnemonic SOHCAHTOA so they can do trigonometry). But since school mathematics is the foremost source of mathematical information for most members of our society, the way mathematics is enacted in school is the way it is understood and enacted by society at large. And so it appears that M_4T is really much more expansive than just formal mathematics; it necessarily includes cultural mathematics too. In short, M_4T is a four-quadrant affair.

But how can any teacher *know* so much? Luckily, as the episodes from the concept study on circles illustrate, cultural mathematics is emergent. While it may make use of some pre-established results, it arises in the moment and is very context-dependent. No two classes give rise to the same cultural mathematics. We believe that mathematics-for-teaching is more about an open disposition to emergent mathematics than it is about mastery of any specific body of knowledge.

Are Circles Efficient? An Open Way of Being

[VOICES: MOSHE AND FREDDIE]

Given the inspiring results we were getting while engaging cultural mathematics, we began to wonder if we could push the envelope even further. For example, can we engage the realm of the aesthetic in our concept study of circles? To this end, we began with a new generative question:

> *How are circles different from squares? Let's think about adjectives that describe circles and others that describe squares.*

There was a big show of hands. The students described circles as "circular, curvy, perfect, and infinite" and squares as "jaggedy, edgy, and straight."

> *These are rather predictable descriptions. Let's go deeper. For example, which is blue and which is red?*

A quick survey showed that over two thirds of the students thought that circles were blue and squares were red.

OK. Try this then. Which of them is sad?

The students were far less certain about this one, although generally they thought that circles were sadder than squares. We then asked them to get into groups and generate lists of adjectives under the headings "circles" and "squares." We then summarized the results.

Circles	happy, thoughtful, wise, soft, realistic, feel better, warmer, transparent, life affirming, dynamic, sleepy
Squares	heavy, masculine, nerdy, deadly

Teacher: It appears that you associated many positive adjectives with circles and negative ones with squares. Why?
Student: It's because circles are perfect. They cannot be criticized.
Teacher: What makes circles perfect?
Student: They have no beginning and no end.
Teacher: Is it good to be perfect? Are you perfect?

The students thought about this question for a while. After some discussion, they agreed that imperfection is one of the essential traits of being human, and that humans are beautiful in their imperfection.

Teacher: What shape is this classroom?
Students: A rectangle.
Teacher: How about the school?
Students: It's rectangular too.
Teacher: If we feel so good about circles why do we build our school in the shape of a rectangle? Can you imagine what this class would look like if it were in the shape of a circle?

The students clearly disapproved of this suggestion.

Student: It would feel weird.
Teacher: Could you build the entire school in the shape of a circle? Would it be good to do so?

Again, the students clearly disliked the idea.

Student: It would be very inefficient to build a circular school.
Teacher: In what sense?

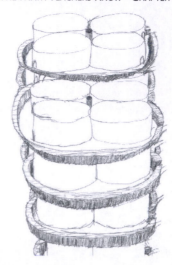

Figure 6.6. A student rendition of a circular school

> *Students: It would not use space well. There would be gaps between the classes. There will be unused space.*
>
> *Teacher: According to this logic, flowers should also be square, in order to maximize efficiency. How many of you would like flowers to be square?*

The students did not think this would be a good idea.

> *Teacher: Who decides what is efficient and what is not? Isn't it a value judgment that we humans impose?*

The class then launched into a conversation about individual and collective values, and how they constrain our choices. We explored how adjectives, such as beautiful or efficient, mediate the ways in which we interact with our environment. Some students who were so inclined then drew renditions of circular schools (e.g., Figure 6.6).

We suspect that some teachers might feel a little uncomfortable with the above episode. Were we not deviating too far afield from "proper" mathematics instruction? Was this really a mathematical engagement worthy of valuable class time? The students certainly had no problem with it. Perhaps they did not make as sharp a distinction between formal and cultural mathematics as some teachers are inclined to do. To us, the episode demonstrates the incredible breadth of cultural mathematics that is lying in wait to be explored in classroom collectives. All that was needed in this case were some generative prompts and openness from us, the teachers, to remain responsive to whatever arose in the discussion.

This open attitude – an open way of being – is a natural consequence of a worldview that conceptualizes mathematics-for-teaching as an emergent phenomenon. It calls for a broad awareness of the dynamic tensions that are at play during each pedagogical encounter with mathematics. Some of these tensions are: formal mathematics *vs.* cultural mathematics, individual knowing *vs.* collective knowing, school mathematics *vs.* grander mathematics, studying for the test *vs.* studying for understanding, explicit knowledge production *vs.* tacit understanding, logical reasoning *vs.* metaphorical thinking. An open way of being entails a willingness to "live in" these tensions dialogically, not privileging either one of their dual ends – in effect, to exchange the tensive "*vs.*" in each of the pairings for the inclusive "**v**". It also entails an openness to the many perspectives through which pedagogical occasions may be interpreted and engaged. An open mindframe considers the best pedagogical responses to be those that promote cultural evolution and life in the classroom.

Our concept study on circles has shown that concept study sensibilities can be at least as potent in the classroom as they are in teachers' professional development. If anything, we have found the 8th-grade students even more responsive to open exploration of cultural mathematics than some practicing teachers. We can see that our original concern about the collective mathematical knowledge inherent in a group of middle-school students was misguided. While it is true that this group of students did not possess extensive knowledge of formal mathematics, it did embody and enact emergent cultural mathematics at every turn, from counting sides of a circle, to stretching ropes around the equator, to the efficiency of circles.

There were no quizzes or tests, and no one felt the poorer in their absence. The students used the formulas for area and circumference of a circle frequently in their investigations. We saw no need to keep reinforcing them with additional exercises. Instead, we asked students to work individually on projects called, "What is interesting for me about circles?" One of the students studied circles in Shakespearean drama. Another studied vortices in black holes. Several students have been experimenting with subtractive fashion design, in which a one-piece garment is created by cutting circular holes into a large piece of cloth. We believe that these sessions opened up a lifelong conversation with circles for the students.

It was very exciting for us to be the teachers in these sessions. We felt that we were able to express ourselves much more fully than in typical classroom situations. We were more authentic. When we did not have answers to some questions, we trusted in the class's collective intelligence and opened them up for group consideration. Progress through the sessions was not linear and we often got stuck at dead ends. Yet each session held new surprises that kept us marveling for days thereafter.

Where Are We Going with This?

Lest it seems that we have lapsed into some rhapsodic, utopian world of mathematics teaching, we close this chapter by re-emphasizing our current understanding of teachers' disciplinary knowledge of mathematics: *M_4T is a way of being with mathematics knowledge that enables a teacher to structure learning situations, interpret student actions mindfully, and respond flexibly, in ways that enable learners to extend understandings and expand the range of their interpretive possibilities through access to powerful connections and appropriate practice.*

To our thinking, Freddie's teaching is a splendid exemplar of this definition. It illustrates not only the teacher's profound mathematical understanding, but also his students' resulting achievement in and appreciation of mathematics. This episode demonstrates the vital link between M_4T and its potential to impact students' lives productively.

For us, Freddie possesses and embodies profound understanding of emergent mathematics. He is always open to emergence in his class, as he mindfully weaves novel mathematics with established results in all four quadrants. We now move to our final chapter in which we define the construct of "profound understanding of emergent mathematics" and provide a glimpse of possible horizons for future research.

THE MATHEMATICS
TEACHERS (NEED TO) KNOW

PROFOUND UNDERSTANDING OF EMERGENT MATHEMATICS

In brief …

We close our discussion of the methodology of concept study by offering the construct of "profound understanding of emergent mathematics" in response to our M_4T question. We then scan the horizons of research and practice for possibilities and needs.

Zero and Function

[VOICES: BRENT AND MOSHE]

In the summer of 2011, a 2-year M_4T master's program patterned after the UBC model was begun in Calgary. One year into their program, the 11 members of the Calgary cohort gathered in the summer of 2012 for an intense, weeklong course entitled "Concept Study."

Having spent a year in classes together, this group was already familiar with much of what had transpired with the UBC group. Indeed, several of the publications based on the Vancouver research had been included as required readings in earlier courses. That meant that participants arrived to the course knowing about realizations, landscapes, entailments, blends, and pedagogical problem solving.

Up to that point in time, however, the group had not engaged in a full-blown concept study. Unlike the UBC cohort, which met face-to-face through almost all of their program experiences, most of the Calgary group's coursework and interactions had been mediated through online environments. As with the UBC group, an attempt had been made during the course on "Research Methods in Education" (Fall Term, 2011) to engage in a preliminary concept study. The topic selected was number decomposition.

Participants did manage to make some significant progress in substructing various aspects number decomposition. However, as they emphasized when they finally gathered in person during the Concept Study course, the missing element of face-to-face interactions contributed to a range of frustrations, most of which revolved around the difficulty of establishing and maintaining a collective knowledge-producing system when agents in that system are not together. As one member described it, participants were "unable to read one another" in online environments – meaning that it seemed impossible to establish the sort of ready ebb and flow that are afforded by immediate, in-person engagements. They shared our expectation that things would be different during the summer course.

By the time that course began, participants were very familiar with details on previous concepts studies of multiplication, *given that topic has served as the focus in most of our reporting to date. As such, we elected to open the course by engaging in a "warm-up" study of* zero. *The topic was chosen because a different group of teachers had just wrapped up a three-month study, and so we had some inkling of where things might go and how we might support this cohort's work.*

Two afternoons were given to the concept study, and a recursive movement through the first four emphases (realizations, landscapes, entailments, and blends) proved to be highly effective. By the end of the second afternoon, the group had arrived at the (meta)realization that while the concept of zero has dozens of not-always-easily-reconciled realizations through the K–12 experience, it emerges through three major elaborations, which they named the "counting zero," the "measuring zero," and the "systemic zero." To truncate the group's extensive discussions, the group developed the following consolidated meanings:

- counting zero – *principal interpretations of "nothing" or "absence," prevalent in the early grades;*
- measuring zero – *principally serving as an orienting or starting value, associated with location- and movement-based interpretations of number, and prevalent in the middle grades; and*
- systemic zero – *arising in algebra and other high school applications, principally serving as signal of an important transition in an object under scrutiny (e.g., the "zero of a function").*

Importantly, these meanings were not seen as distinct, but as emergent – that is, each transcending but including previous realizations. To highlight this insight, the group offered the nested image presented in Figure 7.1.

Despite the fact that it took considerable time, conversation, and argument to arrive at this formulation, there was an interesting collective response when the nested image was finally drawn on the whiteboard. Voiced by one participant, and meeting with nods from around the room: "We really already knew that."

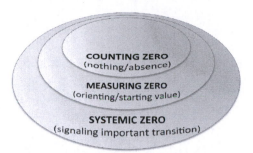

Figure 7.1. A nested metarealization of an emergent-zero concept

In a sense, of course, this knowledge was indeed already present in the group prior to the study. However, it was distributed across the members in a manner that made it tacitly present but explicitly accessible to none. Phrased differently, System 1 rose to claim the insight almost as soon as System 2 reached it.

The following day the group began a second concept study on the topic of function. *The study did not start out smoothly. Because the concept is covered explicitly only in high school, most of the primary- and middle-school teachers in the group were at first a little at sea as mentions of "black box," "input/output," "vertical line test," and so on were offered by their high-school counterparts.*

Things became even more sluggish as the group moved to the landscape emphasis. In particular, given that there is no explicit mention of function prior to high school, the group struggled to find connections between early, middle, and senior grades. Nevertheless, participants persevered – enabled in large part by the insistence of one lower-grades teacher that higher-grades counterparts explain in detail every *one of their realizations so that connections to foci at the elementary level might be identified. It took hours, but eventually the group created a landscape that revealed a flow across three major topics of study: pattern, equation, and function. Once again, the group elected to illustrate the insight as a nested, emergent flow (see Figure 7.2).*

With this work accomplished, movement through the entailments of different realizations (see Figure 7.3) was considerably smoother, setting things up for the activity of collecting the different realizations and their entailments into a metarepresentation that was meaningful and useful to all participants. The final

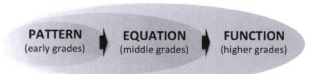

Figure 7.2. A depiction of the nested emergence of the function concept

If a function is defined in terms of ...	... then the type of data is ...	... then it is represented as ...	... the in elementary school it looks like ...	... then in middle school looks like ...	... then in high school it looks like ...	... then the representation type is ...	Conceptual Limitation(s)
vertical line test	discrete/**continuous**	graph	n/a	n/a	graph	enactive and **iconic**	rejects some functions when $x = f(y)$; includes others that are not function
black box	discrete	two related sets of data	missing values in an equation	missing equation problems	two sets of data	iconic	"magical mystery"
input/output	discrete	ordered pairs	sequenced practice (e.g., skip counting)	using linear equations to find missing values; ordered pairs	ordered pairs	iconic and **symbolic**	piecemeal production; missing the whole
independent/dependent relationship	discrete/**continuous**	rule	pattern rules	linear equations; graphing	rule	symbolic	unidirectional: may suggest they're not reversible
1-to-1 relationship/correspondence	discrete/**continuous**	anything in this column	counting based on 1-to-1 correspondence	ordered pairs; T-chart	anything in this column	enactive, iconic, and/or symbolic	implication of uniqueness – i.e., for every x there is a unique y
manipulating/responding variables	discrete/**continuous**	physical event → collect data → graph → interpret	experimental data – growth chart	data; graphing; interpolating/extrapolating tasks	event → data → graph → interpolate/extrapolate	enactive → iconic → **symbolic**	contrived, controlled, contextual
equation with 2 or more variables	discrete/**continuous**	equations (implicit: $y = ...$)	multiple representation problems	equations (implicit: $y = ...$)	equations (implicit: $y = ...$)	symbolic	Some equations with 2 or more variables may not be functions (e.g., $y = \pm\sqrt{x}$)
indexed pattern	discrete	sequence	growing patterns	sequences	sequences	enactive, iconic, and/or **symbolic**	implication of uniqueness; for every i, there is a unique n; dependence on initial position, or starting point; requires a starting point; domain is always natural numbers
mapping	discrete	arrow diagram	classification	transformational arrows	arrow diagram	**iconic** and symbolic	implied container schema
creating a table of values	discrete	horizontal or vertical T-chart	T-chart	T-chart	T-chart	iconic and symbolic	orientation of data; idea of discreteness; possible for left column to be random; implication of uniqueness (all of the limitations?)
$f(x)$	discrete/**continuous**	equations (explicit: $f(x) = ...$)	n/a	equations (implicit: $y = ...$)	equations (explicit: $f(x) = ...$)	symbolic	notation implies multiplication: coordinates are $(x, f(x))$ they don't get that $f(x)$ is y

Figure 7.3. An entailments chart for the concept of *function*

column of the entailments chart, which addresses some of the conceptual limitations of the varied realizations, turned out to be an important innovation to this emphasis. In the group context, it afforded an opportunity to make sense of what a function is by making it clear what a function isn't.

Following that discussion, the suggestion was made that each person should write out their own answer to the question, "What is a function?" on a sheet of paper and then post their responses on the wall. When that was done, the group proceeded to cluster responses according to core theme, and three emerged: function as RELATIONSHIP, function as PATH, and function as VALUE-GENERATOR.

Three subgroups then formed around these themes, and each took on the task of pulling together a cluster of thematically similar interpretations into a consolidated description of function. The three consolidated descriptions were then contrasted. The group undertook another meta-consolidation effort, soon arriving at:

> *A function is a relationship (articulated as a mapping rule, ordered pairings, axes-associating curve, transformation process, etc.) between two sets of values, in which each value in the source set ("the domain") corresponds to at most one value in the target set ("the range").*

Or, in the next, more succinct iteration –

> Function – *a relationship* between two sets of values, in which each value in the source set corresponds to at most one value in the target set.*
>
> * *(articulated as a pattern, equation, mapping rule, ordered pairings, axes-associating curve, transformation process, etc.)*

There are three important details to flag about this emergent realization. First, notice the way in which teachers still acknowledged the role of distinct meanings in their footnote, but displaced the particularity of each meaning from the more distilled definition. That separation-and-connection signals an important difference between the disciplinary knowledge of mathematics teachers and the disciplinary knowledge of mathematicians. Teachers must be more attentive to the nuances of varied meanings, while research mathematicians are more oriented to the formulation of logically sound and more encompassing definitions.

Second, unlike much of mathematics instruction, this definition was derived from meanings – not the other way around. Participants arrived it by consolidating what they knew, not by attempting to make sense of someone else's formulation.

Third, it's also important to note that this emergent definition was not assembled for dissemination. It was a product of the group, intended for the group. And meaningful to the group. As one lower-grades teacher commented, "I didn't give a rip about functions when we started. Now not only do I know what they are, I actually care about them. And I'm pretty sure I know what they are."

The balance of the concept study time was given to pedagogical problem solving, with questions spanning patterning, algebra, and functions across the grades. Only a few of the dozen-or-so problems actually focused on functions, but that didn't seem to matter. The group was clearly disposed to substructing whatever concepts seemed to be involved in the problems raised.

We offer this final anecdote to highlight three core elements of this book. We believe that it illustrates:

- the importance of collective action in the production and interpretation of mathematics for teaching;
- the possibility (and necessity) of developing an open disposition toward substructing mathematics; and
- unique qualities of teachers' disciplinary knowledge of mathematics, specifically its profound nature and its emergent character.

On the first point, we cannot emphasize more strongly our conviction that teachers' mathematics can only be understood systemically, as distributed across a vast and evolving network of persons committed to communicating established and emergent mathematical insights. Contemporary research emphases on identifying and measuring what *individual* teachers can *explicitly* articulate are, in our view, simply inadequate – both as tools to assess what teachers really know and as means to support the development of the vibrant body of M_4T knowledge.[1] To be clear, there are three co-implicated issues here. First, assessments of and supports for teachers' disciplinary knowledge of mathematics tend to be individual-focused and not mindful of the collective. Second, little attention is given to the deeply embodied, tacit, and automatic substrate of teachers' mathematics, relative to the the more explicit dimensions of teacher knowing. Finally, it often feels that much of the discussion of teachers' disciplinary knowledge of mathematics is driven by a deficit model ("What do they lack?") rather than a sufficiency model ("What do they know?").

Pursuant to these issues, and with regard to the second of our bulleted points, above, it has become clear to us that it is indeed possible to nurture an open disposition among teachers toward substructing their mathematics

---

1. To underscore this point, it is interesting to contrast the Calgary cohort's collective response to a research result from Cooney and Wilson (1993). When they asked a similar question, but focused on individuals' explicit knowledge, they found that teachers were typically able to offer only two interpretations of function, namely an equation and a calculation. (See also Harel & Dubinsky, 1992. for similar results.) Even (1998) and Hitt (1998) offer further insight into the emergence of the function concept.

knowledge. Importantly, this work takes time, requires collaboration, and entails some heavy conceptual lifting, but some simple strategies (such as those presented in previous chapters) can greatly support teachers' efforts to better understand what it is they're teaching.

On the subject of our bulleted points, we return to our formulation of the M_4T question …

Q4) How must teachers know mathematics for it to be activated in the moment and in the service of teaching?

… and our preliminary answer …

$A4_1$) M_4T is an open disposition toward mathematics knowledge that enables a teacher to structure learning situations, interpret student actions mindfully, and respond flexibly, in ways that enable learners to extend understandings and expand the range of their interpretive possibilities through access to powerful connections and appropriate practice.

… and now offer a more complete response:

$A4_2$) Teachers must have a *profound understanding of emergent mathematics* (Davis, 2011).

Profound Understanding of Emergent Mathematics

We borrow heavily from the work of others in our formulation of "profound understanding of emergent mathematics," in particular on Liping Ma. In her highly influential and best-selling book, *Knowing and Teaching Elementary Mathematics*, Ma (1999) elaborated Shulman's (1986) notion of "pedagogical content knowledge" to put forward the construct of "profound understanding of fundamental mathematics" (PUFM). The name describes the particular character of teachers' disciplinary knowledge of mathematics as *fundamental* – that is, "foundational, primary, and elementary" (p. 116) – and as *profound* – that is referring to a "deep, vast, and thorough" (p. 120) knowledge of concepts and their interconnections.

PUFM, we believe, reaches across the three questions introduced in Chapter 1:

Q1) What *mathematics* do teachers need to know in order to teach mathematics?
Q2) What *specialized mathematics* (i.e., PCK) do teachers need to know in order to teach mathematics?
Q3) What mathematical knowledge *is entailed by the work* of teaching mathematics?

We resonate with the notion ourselves. However, at the same time we are somewhat uncomfortable with Ma's uses of the adjectives *fundamental* and

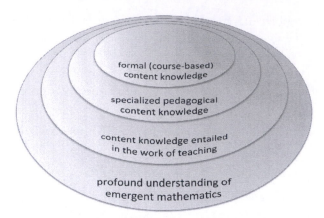

Figure 7.4. Nested aspects of teachers' disciplinary knowledge of mathematics

thorough. Specifically, we would argue that characterizing teachers' disciplinary expertise as "foundational, primary, and elementary" – terms which suggest a closed set of insights and understanding that might be catalogued and assessed — may be antithetical to the project of researching the complexity of teachers' knowledge. As well, the suggestion that such knowledge might be "thorough" for anyone belies its distributed character, its ever-evolving nature, and its vastness.

We thus prefer the notion of "profound understanding of *emergent* mathematics" (PUEM). In brief, we argue that the knowledge needed by teachers is not simply a clear-cut and well-connected set of basics, but a sophisticated, emergent, and largely enactive mix of realizations of mathematical concepts coupled to an awareness of the complex processes through which mathematics is produced. In anchoring our usage of the term *emergent* to the adaptive, evolutionary dynamics described by complexity researchers, we intend to flag the coherent-but-never-fixed character of the complex form of teachers' knowledge. Profound understanding of emergent mathematics is a category of knowing – or, perhaps more accurately, a way of being with – mathematics that includes but elaborates formal content knowledge, specialized pedagogical content knowledge, and the content knowledge entailed in the work of teaching (see Figure 7.4).

Our PUEM construct gestures toward an understanding of teachers' disciplinary knowledge as a responsive and evolving, autopoietic – that is, living and emergent – system of realizations that is distributed across a body of educators. Revisiting the example of multiplication, and on the basis of our analyses of North American textbooks, we feel justified in asserting that most school children will have encountered at least a dozen distinct realizations of multiplication by the 8th grade. It's fair to assume that their teachers are reasonably comfortable with these diverse realizations, and it's probably also reasonable to assume that the diversity is contributing to robust and flexible conceptual understandings (see, e.g., English, 1997; Harel & Confrey, 1994; Lakoff & Núñez, 2000). What may not be reasonable is the expectation that experienced practitioners are conscious of or deliberate with their "choices" as they move fluidly among realizations while teaching. They are able to automatically select and blend interpretations that fit the circumstances at hand, but it's not clear how many are deeply aware what selections have been made, let alone the conceptual entailments of those selections. Even so, the knowledge is there, in the body of teachers.

The webs of association among realizations and concepts, we contend, are not merely accumulative, but aspects of emergent forms. That is, as they interact, they cohere into grander constructs that can open up surprising new dimensions, as exemplified in the concept studies of zero and function of this chapter's opening narrative. In this sense, concepts such as multiplication, zero, and function are elements in an ever-evolving ecosystem of elements, which constitutes the dynamic system of Western mathematics. A profound understanding of emergent mathematics, then, entails both the complex dynamics at work in the development of mathematical knowledge and the specific realizations of elementary concepts that might be relevant and meaningful to learners.

In critiquing Ma's construct we do not intend to dispose of it or diminish its importance. On the contrary, *emergent* mathematics does not preclude what has already been established and embodied in a cultural system. With regard to teachers' knowledge, profound understanding of emergent mathematics includes but broadens Ma's profound understanding of fundamental mathematics. In particular, a profound understanding of emergent mathematics foregrounds and embraces the necessary tensions of stability-and-novelty and coherence-and-decentralization, affording the latter elements of these dyads the same status as the former.

Reiterating the argument at the core of this text, the disciplinary knowledge of mathematics that is needed for effective teaching is more than a set of fundamentals that can be identified, catalogued, transmitted, and tested. It is more than a combination of this knowledge and an ability to unpack concepts for their interconnections, implicit associations, and relevant applications. We believe that M_4T also – and perhaps most critically – entails strong senses of the vibrant, emergent characters of individual learning and the body of mathematical knowledge.

Of course, it's one thing to make this sort of broad claim, and it's quite another to validate it. The field of mathematics education research is rife with reports of small projects but short on large-scale studies of innovations that produced significant and sustainable results. Since we're well aware that our reports also belong to the former category, in the remainder of this closing chapter we will to speculate on what might be needed to move from anecdotal grounding to a strong evidence base.

All That and More …

Teachers' disciplinary knowledge of mathematics is perhaps the most investigated topic among mathematics education researchers today, and is a site of rapid growth of insight. Our intention with this book was not to question or to dismiss any aspect of that work; rather, we have aimed to contribute in a manner that embraces and extends the efforts of our many colleagues who share this research interest. In particular, throughout this writing, we have been oriented by a conviction, shared by many of our colleagues, that teachers and effective teaching matter (cf., Hattie, 2009).

We believe that a vital characteristic of highly effective teachers is strong disciplinary knowledge but, as research has proven, this knowledge cannot be construed using simplistic means, such as a count of college mathematics credits or a written inventory. Rather, M_4T seems to be a knowledge that is available and is enacted in the moment of teaching. Returning to the earlier observation of Baumert and colleagues (2010), there is a relative consensus in the field that the real issue is that teachers' formal disciplinary knowledge of mathematics often remains "inert." Research should therefore focus on the ways in which one's existing disciplinary knowledge might be activated and utilized in pedagogical moments. Our work over the past decade has convinced us that concept study is one means that can be highly effective in this regard.

Again, we acknowledge that we lack the extensive empirical base to make a strong claim in this regard. At the same time, mathematics education researchers have generated an abundance of evidence that, when it comes to school mathematics, many current entrenched methods and emphases are ineffective. Indeed, one can argue that it would be difficult to do any worse. And so, as we consider possible next steps to advance the work presented in this book, we are guided by the conviction that the mathematics education community should not wait until new findings are *proven* before it makes serious efforts to improve on teachers' disciplinary knowledge of mathematics. As a knowledge-producing system with a deeply ethical mandate, the community has an obligation to work together to explore, test, and create new possibilities. Our hope is that the ideas presented herein might trigger some of these explorations.

We believe that much could be gained if mathematics teaching and teacher education included concept-study-like emphases. To that end, there is considerable incidental evidence to make the case for a core component of concept study – that is, for explicating the diversity of interpretations of mathematical concepts. For example, the video portion of the Trends in International Mathematics and Science Study revealed that teachers in Hong Kong and Japan were roughly twice as likely as U.S. teachers to invoke diverse interpretations of concepts (see Hiebert et al., 2003).

Our own research into classroom teaching resources reveals that textbooks used in other parts of the world tend to offer much greater variety of meanings. For example, in an exercise for six-year-olds in one Russian workbook (Peterson, 1996), children are presented with three distinct image-based realizations of addition: combining distances, combining weights, and combining volumes. Significantly, these realizations of addition are much more varied and nuanced than those used in most North American primary texts, where the focus in 1st grade is more typically on interpretations such as gathering objects and counting money – that is, limited to discrete applications and relying only on a NUMBER-AS-COUNT metaphor.

The potential educational significance of this contrast is suggested by studies on the importance of diverse interpretations and associated metaphorical reasoning (Gick & Holyoak, 1983; Richland, Morrison, & Holyoak, 2006; Zook, 1991). In addition to further supporting the call for increased classroom emphasis on variety of realizations, these studies have shown that associative learning can be useful if it is made explicit, and if novices are provided with sustained interpretive assistance.

While it is very likely that teachers' capacity to summon, explicate, and utilize usually-tacit interpretations matters, much work remains to be done on distinguishing among the many possible images and interpretations that might be invoked, on the basis of their relative power and utility. Other questions that could be used to produce a more robust evidence base include:

- How might PUEM as an open disposition be assessed?
- How much of an impact might this have?
- Can concept-study-like emphases be scaled up? If so, how?
- How might such results be operationalized within pre-service and in-service teacher education opportunities?

We offer a few comments on each of these questions.

How might PUEM as an open disposition be assessed?

The assess-ability of PUEM has proven an issue in our efforts to study it. How does one investigate a phenomenon that not only changes constantly, but that changes, in part because of efforts to study it?

To further complicate matters, it's clear that teachers' dispositions toward substructing mathematics can become more open through appropriate engagements. Indeed, nurturing such openness is the principal motivator for developing the concept study methodology. Effects are not necessarily immediate, and as might be expected impacts vary massively from one person to the next. As well, because transformations can be subtle and spread out over considerable time, tracking dispositional changes can be very difficult. Nevertheless, it's hard to argue against the likelihood of important shifts in attitude when teachers are invited to consider together the complexities of learning, of mathematics, and of learning mathematics.

On that point, a question that we must continue to grapple with is how one might access and assess not only the openness of teachers' dispositions, but the associated relevant mathematical knowledge that tends to remain tacit. To be honest, we find this matter perplexing. As noted earlier, our first efforts at interviews and in-class observations met with limited success. Where we have had encouraging results is in monitoring collegial conversations among teachers engaged in concept studies, but such interactions are difficult to track and even more difficult to probe (i.e., in the manner one might query during a clinical interview). In brief, then, its tacit character and its distributed nature, we suspect, are two features of PUEM that researchers must embrace in order to get much traction in studies of the importance of teacher disposition.

How much of an impact might PUEM have?

Of course, the principal issue for us around teachers' mathematics knowledge is the extent to which their PUEM correlates with their effectiveness as mathematics educators.

So far we have accumulated hundreds of anecdotes attesting to the possible significance. We also have some data, gathered on a class-by-class basis, that suggests strong impacts on student attitudes and significant effects on student achievement. As well, at present, we are engaged in a longitudinal study of the impact that teachers' sustained engagement in concept-study activities has on their teaching practice and on their students' understanding. This work is proceeding through focus-group interviews, in-class observations and collaborations, ongoing teacher journaling, and student reporting. We assess student understanding through performance on standardized tests, students' engagement with in-class tasks, students' dispositions toward the discipline, and interview-based evaluations of students' capacity to apply and extend concepts in novel situations.

Intriguing as the emerging data are, they fall far short of the sort of longitudinal and cross-context evidence that is needed to deeply understand the situation and to make strong claims about educational practice.

Moving on the assumption that it will be possible to develop a wider variety of rich, nuanced assessments of teachers' knowledge that capture some of its dispositional, tacit, and distributed elements, the next challenge will be to examine relationships between these aspects and attributes of mathematics teaching and the manners in which their students engage with mathematics.

On that count, we must express our reservation around making heavy use of achievement test results as a strong – or even a good – indicator of impact. For us, more compelling data would include, for example, improved attitudes of students toward mathematics, the emergence of mathematics as a topic of conversation within the student body, the presence of mathematics on students' mathematical horizons, the pursuit of studies and careers in mathematics-reliant fields, and so on.

This matter hits close to home for us. Our home jurisdiction of the province of Alberta, Canada, has done comparatively well on national and international achievement tests in mathematics. Even so, enrollment in mathematics-related university programs has been on a steady decline for decades. Great scores on standardized tests at the end of high school are important, but such scores constitute a wholly inadequate metric when considering the true impact of mathematics teaching. As important as personal achievement is, if the impacts of teaching are not registering on the social and cultural levels, we question their effectiveness.

Can concept-study-like emphases be scaled up? If so, how?

In his meta-analysis of educational research, Hattie (2009) noted that virtually every innovative educational intervention registers an effect. He offered several reasons, including the simple fact that paying attention to one or more details of their practice will instill a renewed mindfulness in teachers – and the simple presence of something novel to prompt that mindfulness might be what really matters (rather than the actual intervention).

As Hattie argues, an effective way to prove that the intervention makes a difference is to scale up. If a strategy works consistently across contexts and for significant numbers of people, it's probably the strategy itself that matters. Unfortunately, efforts in this direction can easily get caught in a Catch-22 situation within social sciences research, given the impossibility of providing the extensive sorts of empirical evidence that are needed to justify scaling up in advance of the actual scaling.

Assuming that such quagmires might be traversed or sidestepped, the "simple" matter of scaling up presents its own set of complex issues. For example, how does one secure the resources necessary to move from a handful of people to a small army? If such scaling is actually accomplished, how might one ensure fidelity of the intervention? As a research community, mathematics educators are still far from making definitive claims about the relationships between teachers' profound understandings of mathematics and their students' mathematical understandings. Our suspicion is that efforts to address this vexing quandary will require more fine-grained analyses than large-scale assessments, in large part because many of the most important aspects of teachers' knowledge are simply unavailable for explicit and immediate assessment. They are tacit and can only emerge through participation in collective explorations, such as concept studies.

That said, we have engaged in some pragmatic strategies to ramp up our population base. For example, the teacher education program at the University of Calgary has been restructured so that *all* teacher candidates – at both elementary and secondary levels – must declare a specialization. For those who opt for a mathematics specialization, a major component of their two-year (four-semester) experience is concept study. As well, the Faculty of Education is offering both a two-year Master's of Education cohort (based on the one described in Chapters 4 to 6) and a four-course graduate certificate that emphasize concept study.

Those efforts, of course, focus on individuals. Another strategy is to concentrate on collective levels, such as schools and school districts. One project of this sort is just commencing, involving most of the teachers who deal with mathematics in a K–12-school in Calgary. Once again organized around concept study, this five-year project is looking at the possible impact on the culture of the school. That is, the "learner" in this case, is seen as the community, and the metrics include increased engagement in mathematics and the emergence of mathematics-related fields on graduates' career horizons.

Other, as-yet-unengaged sites of emphasis include collaborations between faculties of education and departments of mathematics (particularly around required mathematics courses), collaboration with other schools of education, and the use of social media to create and support on-line M_4T communities. To our minds, an exciting side benefit of this sort of imaginings is the possible emergence of opportunities to examine simultaneously both the *system of knowledge produced* and the *systems producing the knowledge*. There would seem great potential of such studies for deepening collective knowledge of, for example, how teachers and students co-construct understanding and generate new insights.

How might such results be operationalized within pre-service and in-service teacher education opportunities?

Assuming that significant progress might be made on better understanding

PUEM and its relationship to student learning, the immediate follow-up question for us is how insights and findings might be translated into effective structures for pre-service and in-service teacher education. We are working on precisely these matters now through a few projects that combine pre-service in in-service teacher education in the context of Partner Research Schools, and we expect to have gathered sufficient data to begin reporting our findings within a few years.

In the meantime, many subquestions are presenting themselves. How, for example, might extended engagement in concept study compare to the effects of extended engagement in lesson study? What differences does it make to drill deep into a concept rather than to focus on a structure for presenting an aspect of a concept? How might an emphasis on teachers' profound understanding of *emergent* mathematics produce different effects to an emphasis on teachers' profound understanding of *fundamental* mathematics? What difference might it make to overlay conventional studies of teachers' disciplinary knowledge with an engaged emphasis on the complex, dynamic, participatory nature of knowledge production?

We hasten to re-emphasize here that we do not intend a this-or-that sensibility as we contemplate these sorts of questions. As developed in Chapter 1, the notion of "profound knowledge of emergent mathematics," embraces the many insights gleaned by researchers who posed the question of mathematics teachers' disciplinary knowledge in different ways. We are not wondering how others might be wrong or how they have fallen short. Rather, we want to know more about how the different research attitudes might complement one another and how they might open up a rich conversation.

Where Are We Going with This?

One of the more intriguing results to emerge from linguistics and the cognitive sciences in the past quarter century is the recognition that the most important influencers of language are *not* adults, but children. Children embrace new vocabulary and smooth out inconsistencies of grammar more readily than adults. They are not as bound by conventions and rules that have become invisible to expert users of language – that is, adults (Deacon, 1997). This is not to say that adults are without influence. However, when the sites and sources of adult influence are examined closely, it turns out that the adults who yield the greatest influence are those that interact most extensively with children. These are teachers. They work with dozens of children, day in and day out, negotiating conventions, co-constructing meaning, co-defining the importance of elements of language.

To our knowledge, researchers lack the detailed evidence to make the same claims about mathematics, but we are nonetheless intrigued by the suggestion that students and their teachers may be the most potent shapers of mathematics. For the vast majority of the world's population, the school classroom is the principal interface of formal mathematics knowledge. It is there that what is mathematically thinkable, interpretable, and applicable are defined, practiced, and assessed. The impact manifests not only on the personal level. School mathematics also defines cultural possibility. As an illustration of this point, consider this simple thought experiment: imagine what would happen if computer algebra systems were to replace current pencil-based algebra in schools.

Among the consequences, it is likely that students would have access to a powerful tool that allows them to model and solve a variety of very complex real-life problems. It's easy to imagine that, within one or two decades, society as a whole would gain a very different perspective on the possibilities that inhere in the discipline.

If teachers are the primary shapers of mathematics in society, then the structures and emphases that have defined teachers' disciplinary expertise must be challenged and reframed. We thus end by clarifying how we understand the relationship between *mathematics* and *mathematics-for-teaching*.

The principal goal of concept study is to create new possibilities for mathematics teaching by exposing teachers to more nuanced understandings and elaborations of extant mathematics. The goal is not to create new formal mathematics, a task that would require very different validation criteria. The essential questions for us do not revolve around the ontological status of mathematical concepts or around teachers' production of new mathematics. Rather, we seek to study new emergent possibilities for *understanding* mathematics. This framing is consistent with the recognition that effective teaching is not simply a matter of transmission, but always entails transformation. While discussions of transformation usually revolve around the transformation of learners, we believe, as do others (e.g., Brousseau, 1997), that mathematical knowledge also resides within the realm of teachers' influence. After all, mathematics teachers usually have the most direct and pervasive influence on defining what is mathematically interpretable for most of the population.

Lest we leave off on such an elevated plane, we remind ourselves that our driving concern, throughout our careers, has been with making mathematics accessible to learners. And, in our teaching experience, access to mathematical understanding is gained much more readily through interpretation than through definition or procedure. All too often we have observed learners who were faced with sudden conceptual complication and little interpretive possibility or assistance. More often then not their response has been disaffection and reduced interest in the subject matter. Alternatives to profound understanding and open disposition – i.e., rote memorization and routinized application – make for unengaging mathematics that is ill suited to today's emerging needs.

Current studies on teachers' disciplinary knowledge appear to be coevolving with perspectives on why mathematics is taught in the first place. Schools have traditionally emphasized the development of technical competence, which was an obvious need in an industrial economy. But in a knowledge-based economy, the development of conceptual fluency is of greater importance and has been the focus of major initiatives in school mathematics. Our research into the subtlety and complexity of teachers' knowledge not only reveals that some major shortcomings of these initiatives, it also offers an important possible route to achieving the goal of true conceptual fluency – a profound understanding of emergent mathematics.

REFERENCES

Adler, J., & Davis, Z. (2006) Opening another black box: research mathematics for teaching in mathematics teacher education. *Journal for Research in Mathematics Education, 37*(4), 270–296.

Ainsworth, S. (1999) The functions of multiple representations. *Computers & Education, 33*, 131–152.

Ainsworth, S., & van Lebeke, N. (2004) Multiple forms of dynamic representation. *Learning and Instruction, 14*, 235–357.

Arnold, V.I. (1997 March 7) *Discussion on the teaching of mathematics.* Paris, FR: Palais de Découverte.

Baker, S. (2008) *The numerati.* Boston: Houghton Mifflin.

Ball, D.L., & Bass, H. (2000) Interweaving content and pedagogy in teaching and learning to teach: Knowing and using mathematics. In J. Boaler (Ed.), *Multiple perspectives on the teaching and learning of mathematics* (pp. 83–104). Westport, CT: Ablex.

Ball, D.L., & Bass, H. (2003) Toward a practice-based theory of mathematical knowledge for teaching. In E. Simmt & B. Davis (Eds), *Proceedings of the 2002 Annual Meeting of the Canadian Mathematics Education Study Group* (pp. 3–14). Edmonton, AB: CMESG/GCEDM.

Ball, D.L., Blömkeke, S., Delaney, S., & Kaiser, G. (Eds) (2102) Measuring teacher knowledge – approaches and reselts from a cross-national perspective. Special issue of *ZDM: The International Journal of Mathematics Education, 44*(3).

Ball, D.L., Hill, H.C., & Bass, H. (2005) Knowing mathematics for teaching: Who knows mathematics well enough to teach third grade, and how can we decide? *American Educator,* Fall, 14–46.

Ball, D.L., Lubienski, S.T., & Mewborn, D.S. (2001) Research on teaching mathematics: the unsolved problem of teachers' mathematical knowledge.

In V. Richardson (Ed.), *Handbook of research on teaching* (pp. 433–456). Washington, DC: American Educational Research Association.

Ball, D.L., Thames, M.H., & Phelps, G. (2008) Content knowledge for teaching: what makes it special? *Journal of Teacher Education, 59*(5), 389–407.

Bauerserfeld, H. (1992) Integrating theories for mathematics education. *For the Learning of Mathematics, 12*(2), 19–28.

Baumert, J., Kunter, M., Blum, W., Brunner, M., Voss, T., Jordan, A., Klusmann, U., Krauss, S., Neubrand, M., Tsai, Y. (2010) Teachers' mathematical knowledge, cognitive activation in the classroom, and student progress. *American Educational Research Journal*, 47, 133–180.

Begle, E.G. (1972) SMSG Reports No. 9: *Teacher knowledge and student achievement in algebra*. Palo Alto, CA: Stanford University.

Begle, E.G. (1979) *Critical variables in mathematics education: findings from a survey of the empirical literature*. Washington, DC: Mathematical Association of American and National Council of Teachers of Mathematics.

Bejan, A., & Zane, J.P. (2012) *Design in nature: how the constructal law governs evolution in biology, physics, technology, and social organization*. New York: Doubleday.

Brousseau, G. (1997) *Theory of didactical situations in mathematics: didactiques des mathématiques, 1970–1990*. (N. Balacheff, M. Cooper, R. Sutherland, & V. Warfield, Trans.). Dordrecht, NL: Kluwer.

Brunner, M., Kunter, M., Krauss, S., Klusmann, U., Baumert, J., Blum, W., et al. (2006) Die professionalle kompetenz von Mathematiklehrkräften: Konzeptualisierung, erfassung und bedeutung für den unterricht. Eine zwischenbilanz des COACTIVE-Projekts. In M. Prenzel & L. Allolio-Näcke (Eds), *Untersuchungen zur Bildungsqualität von Schule* (pp. 54–83). Münster, DE: Waxmann.

Bruner, J. (1966) *Toward a theory of instruction*. Cambridge, MA: Harvard University Press.

Burger, E.B., & Starbird, M. (2005) *The heart of mathematics: an invitation to effective thinking* (2nd edn.). Emeryville, CA: Key College Publishing

Capra, F. (2002) *The hidden connections*. London: HarperCollins.

Chase, W.G., & Simon, H.A. (1973) The mind's eye in chess. In W. G. Chase (Ed.), *Visual information processing* (pp. 215–281). New York: Academic Press.

Cobb, P. (1994) Where is the mind? Constructivist and sociocultural perspectives on mathematical development. *Educational Researcher, 23*(7), 13–20.

Cobb, P. (1999) Individual and collective mathematics development: the case of statistical data analysis. *Mathematics Thinking and Learning, 1*(1), 5–43.

Cobb, P., Wood, T., Yackel, E., & McNeal, B. (1992) Characteristics of classroom mathematics traditions: an interactional analysis. *American Educational Research Journal, 29*, 573–604.

Cooney, T.J., & Wilson, M.R. (1993) Teachers' thinking about functions; historical and research perspectives. In T. Romberg, E. Fennema, & T. Carpenter (Eds), *Integrating research on the graphical representation of function* (pp. 131–158). Hillsdale, NJ: Erlbaum.

Davidson, N. (2011) *Now you see it: how the brain science of attention will transform the way we live, work, and learn.* New York: Viking.

Davis, B. (1989) *Meaningful talk in the mathematics classroom.* Unpublished master's thesis. Edmonton, AB: University of Alberta.

Davis, B. (1994) Mathematics teaching: moving from telling to listening. *Journal of Curriculum and Supervision, 9*(3), 267–283.

Davis, B. (1996) *Teaching mathematics: toward a sound alternative.* New York: Garland Publishing.

Davis, B. (1997) Listening for differences: an evolving conception of mathematics teaching. *Journal for Research in Mathematics Education, 28*(3), 355–376.

Davis, B. (2008) Is 1 a prime number? Developing teacher knowledge through concept study. *Mathematics Teaching in the Middle School, 14*(2), 86–91.

Davis, B. (2011) Mathematics teachers' subtle, complex disciplinary knowledge. *Science, 332*, 1506–1507.

Davis, B., & Renert, M. (2009) Mathematics for teaching as shared, dynamic participation. *For the Learning of Mathematics, 29*(3), 37–43.

Davis, B., & Renert, M. (2012) Profound understanding of *emergent* mathematics: broadening the construct of teachers' disciplinary knowledge. *Educational Studies in Mathematics, 82*, 245–265.

Davis, B., & Simmt, E. (2003) Understanding learning systems: Mathematics teaching and complexity science. *Journal for Research in Mathematics Education, 34*(2), 137–167.

Davis, B., & Simmt, E. (2006) Mathematics-for-teaching: an ongoing investigation of the mathematics that teachers (need to) know. *Educational Studies in Mathematics, 61*(3), 293–319.

Davis, B., & Sumara, D. (2006) *Complexity and education: inquiries into learning, teaching, and research.* Mahwah, NJ: Lawrence Erlbaum Associates.

Davis, B., Sumara, D., & Luce-Kapler, R. (2008) *Engaging minds: changing teaching in complex times* (2nd edn.). New York: Routledge.

Davis, P.J., & Hersh, R. (1986) *Descartes' dream: the world according to mathematics.*

Boston: Houghton Mifflin.

Deacon, T.W. (1997) *The symbolic species: the co-evolution of language and the human brain.* New York: W.W. Norton

diSessa, A.A. (2004) Metarepresentation: native competence and targets for instruction. *Cognition and Instruction, 22*(3), 293–331.

Dweck, C.S. (2006) *Mindset: the new psychology of success.* New York: Random House.

English, L. (Ed.) (1997) *Mathematical reasoning: analogies, metaphors, and images* Hillsdale, NJ: Erlbaum.

Ericsson, K.A., Charness, N., Feltovich, P., & Hoffman, R.R., (2006) *Cambridge handbook on expertise and expert performance.* Cambridge, UK: Cambridge University Press.

Even, R. (1998) Factors involved in linking representations of function. *Journal of Mathematical Behavior, 17*(1), 105–121.

Even, R., & Li, Y. (Eds) (2011) Approaches and practices in developing teachers' expertise in mathematics instruction. Special issue of *ZDM: The International Journal of Mathematics Education, 43*(6–7).

Even, R., & Tirosh, D. (2002) Teacher knowledge and understanding of students' mathematical learning. In L.D. English (Ed.), *Handbook of International Research in Mathematics Education* (pp. 219–240). Mahwah, NJ: Lawrence Erlbaum Associates.

Fauconnier, G., & Turner, M. (2003) *The way we think: cnceptual blending and the mind's hidden complexities.* New York: Basic Books.

Fauvel, J., & van Maanen, J. (Eds) (2000) *History in mathematics education: the 10th ICME study.* Dordrecht, NL: Kluwer.

Fernandez, C., & Yoshida, M. (2004) *Lesson study: a Japanese approach to improving mathematics teaching and learning.* Mahwah, NJ: Erlbaum.

Fischbein, E. (1989) Tacit models and matheamtical reasoning. *For the Learning of Mathematics, 9*(2), 9–14.

Fischbein, E. (1993) The theory of figural concepts. *Educational Studies in Mathematics, 24*, 129–162.

Foote, R. (2007) Mathematics and complex systems. *Science, 318*, 410–412.

Freudenthal, H. (1983) *Didactical phenomenology of mathematical structures.* Dordrecht, NL: D. Reidel.

Gibbs, Jr., R.W., & Colston, H.L. (2007) *Irony in language and thought: a cognitive science reader.* New York: Routledge.

Gick, M.L., & Holyoak, K.J. (1983) Schema induction and analogical transfer. *Complexity and Organization, 1*(1), 49–72.

Goldin, G.A., & Janvier, C. (Eds) (1998) Representations and the psychology of mathematics education, Parts I and II (special issues). *Journal of Mathematical Behavior, 17*(1 & 2).

Goldstein, J. (1999) Emergence as a construct: history and issues. *Emergence: Complexity and Organization 1*(1), 49–72.

Gowers, T. (2002) *Mathematics: a very short introduction.* New York: Oxford University Press.

Greer, B. (1994) Extending the meaning of multiplication and division. In G. Harel & J. Confrey (1994), *The development of multiplicative reasoning in the learning of mathematics* (pp. 61–87). Albany, NY: State University of New York Press.

Harel, G., & Confrey, J. (Eds) (1994) *The development of multiplicative reasoning in the learning of mathematics* Albany, NY: State University of New York Press.

Harel, G., & Dubinsky, E. (Eds). (1992) The concept of function; aspects of epistemology and pedagogy. *MAA Notes*, no. 28.

Hattie, J. (2009) *Visible learning: a synthesis of over 800 meta-analyses relating to achievement.* New York: Routledge

Hiebert, J., Gallimore, R., Garnier, H. Bogard Givvin, K. Hollingsworth, H., Jacobs, J., Miu-Ying Chui, A., Wearne, D., Smith, M., Kersting, N., Manaster, A., Tseng, E., Etterbeek, W., Manaster, E., Gonzales, P., & Stigler, J.W. (2003) *Teaching mathematics in seven countries: the results of the TIMSS 1999 Video Study.* Washington, DC: National Center for Education Statistics. Department of Education.

Hill, H., Ball, D.L., & Schilling, S. (2008) Unpacking "pedagogical content knowledge": conceptualizing and measuring teachers' topic-specific knowledge of students. *Journal for Research in Mathematics Education, 39*(4), 372–400.

Hill, H.C., Rowan, B., & Ball, D.L. (2005) Effects of teachers' mathematical knowledge for teaching on student achievement. *American Educational Research Journal, 42*(2), 371–406.

Hitt, F. (1998) Difficulties in the articulation of different representations linked to the concept of function. *Journal of Mathematical Behavior, 17*(1), 123–134.

Janvier, C. (Ed.) (1987) *Problems of representation in the teaching and learning of mathematics.* Hillsdale, NJ: Lawrence Erlbaum Associates.

Jenkins, H., Clinton, K., Purushotma, R., Robison, A. J., & Weigel, M. (2006) *Confronting the challenges of participatory culture: Media education for the 21st century.* Available at www.digitallearning.macfound.org/atf/cf/%7B7E45C7E0-A3E0-4B89-AC9C-E807E1B0AE4E%7D/jenkins_white_paper,pdf

Johnson, S. (2001) *Emergence: the connected lives of ants, brains, cities, and software.* New York: Scribner.

Kahneman, D. (2011) *Thinking, fast and slow*. New York: Farrar, Straus and Giroux.

Kieran, C. (2001) The mathematical discourse of 13-year-old partnered problem solving and its relation to the mathematics that emerges. *Educational Studies in Mathematics, 46*, 187–228.

Kieren, T.E. (2000) Dichotomies or binoculars: reflections on the papers by Steffe and Thompson and by Lerman. *Journal for Research in Mathematics Education, 31*, 228–233.

Kilpatrick, J., Swafford, J., & Findell, B. (2001) *Adding it up: helping children learn mathematics*. Washington, DC: National Academy Press.

Lakoff, G., & Johnson, M. (1999) *Philosophy in the flesh: the embodied mind and its challenge to western thought*. New York: Basic Books.

Lakoff, G., & Núñez, R. (2000) *Where mathematics comes from: how the embodied mind brings mathematics into being*. New York: Basic Books.

Leinhardt, G. (1989) Math lessons: a contrast of novice and expert competence. *Journal for Research in Mathematics Education, 20*(1), 52–75.

Leinhardt, G., Putnam, R., & Hattrup, R.A. (1992) *Analysis of arithmetic for mathematics teaching*. Hillsdale, NJ: Lawrence Erlbaum Associates.

Lesh, R., Post, T., & Behr, M. (1987) Representations and translations among representations in matheamtics learning and problem solving. In C. Janvier (Ed.), *Problems of representation in the teaching and learning of mathematics* (pp. 33–40). Hillsdale, NJ: Lawrence Erlbaum Associates.

Ma, L. (1999) *Knowing and teaching elementary mathematics: teachers' understanding of fundamental mathematics in China and the United States*. Mahwah, NJ: Erlbaum.

Marton, F. & Säljö, R (1976a) On qualitative differences in learning – 1: outcome and process. *British Journal of Educational Psychology, 46*, 4–11.

Marton, F. & Säljö, R (1976b) On qualitative differences in learning – 2: outcome as a function of the learner's conception of the task. *British Journal of Educational Psychology, 46*, 115–127.

Mazur, B. (2003) *Imagining numbers (particularly the square root of minus fifteen)*. New York: Farrar Straus Giroux.

Mewborn, D. (2001) Teachers' content knowledge, teacher education, and their effects on the preparation of elementary teachers in the United States. *Mathematics Education Research Journal, 3*, 28-36.

Mitchell, M. (2009) *Complexity: a guided tour*. Oxford, UK: Oxford University Press.

Monk, D.H. (1994) Subject area preparation of secondary mathematics and science teachers and student achievement. *Economics of Education Review, 13*(2), 125–145.

NCTM (National Council of Teachers of Mathematics). (1989) *Curriculum and evaluation standards for school mathematics.* Reston, VA: NCTM.

NCTM (National Council of Teachers of Mathematics). (2000) *Principles and standards for school mathematics.* Reston, VA: NCTM.

Peterson, L.G. (1996) *Mathematics: grade 1, part 2.* Moscow: Yuventa.

Pirie, S.E.B., & Kieren, T.E. (1994) Growth in mathematical understanding: how can we characterize it and how can we represent it? *Educational Studies in Mathematics, 26,* 165–190.

Polanyi, K.M. (1966) *The tacit dimension.* New York: Doubleday.

Presmeg, N.C. (1986) Visualization in high school mathematics. *For the Learning of Mathematics, 6*(3), 42–46.

Renert, M., & Davis, B. (2010a) An open way of being: Integral reconceptualization of mathematics for teaching. In Esbjörn-Hargens, S., Reams, J. & Gunnlaugson, O. (Eds), *Integral education: New directions for higher learning* (pp.193-215). Albany, NY: SUNY Press

Renert, M., & Davis, B. (2010b) Life in mathematics: evolutionary perspectives on subject matter. In Walshaw, M. (Ed.), *Unpacking pedagogy: new perspectives for mathematics classrooms* (pp. 177-200). Charlotte, NC: Information Age.

Richland, L.E., Morrison, R.G., & Holyoak, K.J. (2006) Children's development of analogical reasoning: Insights from scene analogy problems. *Journal of Experimental Child Psychology, 94,* 249–273.

Rowland, T., & Ruthven, K. (Eds) (2011) *Mathematical knowledge in teaching.* Dordrecht, NL: Springer.

Scardamalia, M., & Bereiter, C. (2003) Knowledge building. In J.W. Guthrie (Ed.), *Encyclopedia of Education.* 2nd edition. New York: Macmillan Reference.

Schoenfeld, A.H. (1985) *Mathematical problem solving.* New York: Academic Press.

Schoenfeld, A.H., Herrmann, D.J. (1982) Problem perception and knowledge structure in expert and novice mathematical problem solvers. *Journal of Experimental Psychology: Learning, Memory, and Cognition, 8*(5), 484–494.

Seufert, T. (2003) Supporting coherence formation in learning from multiple representations. *Learning and Instruction, 13,* 227–237.

Sfard, A. (1991) On the dual nature of mathematical conceptions: reflections on processes and objects as different sides of the same coin. *Educational Studies in Mathematics, 22,* 1–36.

Sfard, A. (2008) *Thinking as communicating: human development, the growth of discourses, and mathematizing.* New York: Cambridge University Press.

Sfard, A., & Kieran, C. (2001) Cognition as communication: rethinking learning-by-talking through multi-faceted analysis of students' mathematical interactions. *Mind, Culture, and Activity, 8*(1), 42–76.

Shulman, L.S. (1986) Those who understand: knowledge growth in teaching. *Educational Researcher, 15*(2), 4–14.

Surowiecki, J. (2004) *The wisdom of crowds.* New York: Doubleday.

Tall, D.O. (2004) Thinking through three worlds of mathematics, *Proceedings of the 28th Conference of the International Group for the Psychology of Mathematics Education,* Bergen, Norway, vol. 4, 281–288.

Usiskin, Z., Peressini, A., Marchisotto, E.A., & Stanley, D. (2003) *Mathematics for high school teachers: an advanced perspective.* Upper Saddle River, NJ: Pearson

Van de Walle, J., Karp, K.S., & Bay-Williams, J.M. (2012) *Elementary and middle school mathematics: teaching developmentally* (8th edn.). Upper Saddle River, NJ: Pearson.

von Glasersfeld, E. (1990) An exposition of constructivism: why some like it radical. In R.B. Davis, C.A. Maher, & N. Noddings (Eds), *Constructivist views on the teaching and learning of mathematics* (pp. 19–29). Reston, Virginia: National Council of Teachers of Mathematics.

Waldrop, M.M. (1992) *Complexity: the emerging science on the edge of order and chaos.* New York: Simon & Schuster.

Wilber, K. (1995) *Sex, ecology, spirituality: The spirit of evolution.* Boston: Shambhala

Wilber, K. (2006) *Integral spirituality: A startling new role for religion in the modern and postmodern world.* Boston: Integral Books.

Wilson, S.M., Floden, R.E., & Ferrini-Mundy, J. (2002) Teacher preparation research: an insider's view from the outside. *Journal of Teacher Education, 53*(3), 190–204.

Zack, V., & Graves, B. (2001) Making mathematical meaning through dialogue: once you think of it, the z minus three seems pretty weird. *Educational Studies in Mathematics, 46,* 229–271.

Zazkis, R. (2011) *Relearning mathematics: a challente for prospective elementary school teachers.* Charlotte, NC: Information Age Publishing.

Zook, K.B. (1991) Effects of analogical processes on learning and misrepresentation. *Educational Psychology Review, 3,* 41–72.

INDEX